80th Annual Eastern Snow Conference 2024

Waterloo, Canada
4-6 June 2024

ISBN: 979-8-3313-2057-7

Printed from e-media with permission by:

Curran Associates, Inc.
57 Morehouse Lane
Red Hook, NY 12571

Some format issues inherent in the e-media version may also appear in this print version.

Printed with permission by Curran Associates, Inc. (2025)

For permission requests, please contact Eastern Snow Conference
at the address below.

Eastern Snow Conference
C/O Dr. Krystopher Chutko
117 Science Pl-Dept. Geography
Saskatoon, Sk, Canada S7N 5C8

https://www.easternsnow.org/

Additional copies of this publication are available from:

Curran Associates, Inc.
57 Morehouse Lane
Red Hook, NY 12571 USA
Phone: 845-758-0400
Fax: 845-758-2633
Email: curran@proceedings.com
Web: www.proceedings.com

Proceedings of the
80th ANNUAL

EASTERN SNOW CONFERENCE

4-6 June 2024

University of Waterloo, Waterloo, Ontario, Canada

ISSN: 0424-1932

Proceedings of the 80th Annual Eastern Snow Conference

Hilda Snelling
1932-2024

The Proceedings of the 80th Annual Eastern Snow Conference is dedicated to Air Force Captain, meteorologist, and Eastern Snow Conference Life Member Hilda Snelling, who passed away on April 1, 2024 in New Port Richey, Florida, USA, at the age of 91.

CONTENTS

Oral Session #1

Poster Session #1

Oral Session #2

Oral Session #3

Poster Session #2

Oral Session #4

Oral Session #5

Oral Session #6

FOREWORD

This proceedings volume contains papers presented at the 80th Eastern Snow Conference (ESC) held 4-6 June 2024 at the University of Waterloo in Waterloo, Ontario. The meeting featured oral and poster presentations covering a wide array of cryospheric science, including but not limited to the observation of High Arctic glacier mass balance, the application of several remote sensing techniques to the measurement of snow cover, snowmelt contributions to streamflow, and the development and evaluation of snow and ice monitoring sensors.

The ESC is a joint United States and Canadian forum for discussing recent work on operational, applied and scientific issues related to snow and ice. It also retains an increasing interest as a symposium where novel approaches to cryospheric science of international significance are presented. The ESC has published an annual series of proceedings since 1952. Typical topics include studies of snow and ice as materials, snow removal, meteorological forecasting, river ice control, snow hydrology, snow chemistry, glaciology, remote sensing of snow and ice, and snow ecology. Membership in the ESC is open to all interested individuals and corporations. Additional copies of the current proceedings and all back issues can be obtained from the Secretary. Additional information about the Eastern Snow Conference may be found at: http://www.easternsnow.org/.

We thank members of the ESC executive committee for their continued work toward making each annual meeting a success. This includes, but is not limited to, the work of the Research Committee who evaluate all student presentations and recommend awardees for the three student prizes.

The 2024 meeting of the Eastern Snow Conference and these Proceedings were made possible in part by Corporate membership from Campbell Scientific Canada, Geonor, and Hoskin Scientific Ltd.

We look forward to seeing you at the 81st Eastern Snow Conference meeting to be held 10-12 June 2025 at Syracuse University in Syracuse, New York, USA.

Krystopher Chutko and Joan Ramage
ESC Proceedings Co-Editors
University of Saskatchewan and Environment and Lehigh University

STATEMENT OF PURPOSE

The Eastern Snow Conference (ESC) is a joint Canadian/US organization founded in the 1940s, originally with members primarily from eastern North America. Our current members are scientists, snow surveyors, engineers, technicians, professors, students, and operational and maintenance professionals from around the world. There is a western counterpart to the ESC, the Western Snow Conference (WSC), which also is a joint Canadian/US organization.

The Eastern Snow Conference is a forum that brings the research and operations communities together to discuss recent work on scientific, applied, and operational issues related to snow and ice. The location of the conference alternates yearly between the United States and Canada, and attendees present their work by either giving a talk or presenting a poster. Most resulting papers are reviewed, edited, and published in our yearly *Proceedings of the Eastern Snow Conference*. In recent years, the ESC meetings have included sessions on snow physics, winter survival of animals, snow and ice loads on structures, river ice, remote sensing of snow and ice, and glacier processes. Volumes of the *Proceedings* can be found in libraries throughout North America and Europe, and the papers are also available through the National Technical Information Service (NTIS) in the United States and CISTI in Canada. All volumes of the Proceedings are now available at *easternsnow.org*.

* * * * * * * * *

Le Colloque sur la neige région Est (ESC) est une organisation américaine-canadienne fondée dans les années '40 et dont les membres provenaient à l'origine principalement de l'est de l'Amérique-du-Nord. Actuellement, les membres, qu'ils soient chercheurs, techniciens, ingénieurs, professeurs, étudiants, et spécialistes des services d'exploitation et d'entretien, viennent non seulement d'Amérique-du-Nord, mais aussi du Royaume-Uni, du Japon, et d'Allemagne. Le Colloque sur la neige-région Ouest (WSC), aussi une organisation américaine-canadienne, est l'homologue de l'ESC pour l'ouest nord-américain.

L'ESC est un forum qui rassemble chercheurs et responsables des services d'exploitation pour discuter des travaux récents sur les problèmes scientifiques, opérationnels, ou autres dus à la neige et à la glace. Le site de cette réunion annuelle alterne entre les États Unis et le Canada. Les participants y présentent les résultats de leurs travaux par des communications orales ou au moyen d'affiches. Ces communications, une fois revues et éditées, sont publiées dans les *Annales de l'ESC*. Dans les années récentes, les réunions de l'ESC ont inclus des sessions sur la physique de la neige, la survie hivernale de la faune, les forces exercées par la neige et la glace sur les structures et les bâtiments, la glace de rivière, la télédétection de la neige et de la glace, et les processus glaciaires. Les *Annales de l'ESC* sont accessibles dans la plupart des bibliothèques scientifiques d'Amérique-du-Nord et d'Europe. Des copies d'articles peuvent être obtenues du National Technical Information Service (NTIS) aux États Unis et son équivalent au Canada, le CISTI. Toutes les volumes des *Annales de l'ESC* sont maintenant disponibles sur *easternsnow.org*.

EXECUTIVES FOR THE 2023-24 80th EASTERN SNOW CONFERENCE

PRESIDENT

Joan Ramage
Bethlehem
Pennsylvania, USA

PAST PRESIDENT

Craig Smith
Saskatoon
Saskatchewan, Canada

VICE-PRESIDENT AND PROGRAM CHAIR

Grant Gunn
Waterloo
Ontario, Canada

SECRETARY-TREASURER (Canada)

Krystopher Chutko
Saskatoon
Saskatchewan, Canada

SECRETARY-TREASURER (USA)

Eli Deeb
Hanover
New Hampshire, USA

ESC PROCEEDINGS EDITORS

Krystopher Chutko
Saskatoon
Saskatchewan, Canada

Joan Ramage
Bethlehem
Pennsylvania, USA

***PHYSICAL GEOGRAPHY* EDITORS**

Mauri Pelto
Dudley
Massachusetts, USA

STEERING COMMITTEE

Laura Thomson (Chair)
Kingston
Ontario, Canada

Sean Helfrich
Suitland
Maryland, USA

Carrie Vuyovich
Hanover
New Hampshire, USA

James Brylawski
Augusta
New Jersey, USA

David Robinson
New Brunswick
New Jersey, USA

RESEARCH COMMITTEE

Sam Tuttle (Chair)
Syracuse
New York, USA

Melissa Wrzesien
Greenbelt
Maryland, USA

Eunsang Cho
Greenbelt
Maryland, USA

Tara Troy
Victoria
British Columbia, Canada

Fraser King
Waterloo
Ontario, Canada

Benoit Montpetit
Ottawa
Ontario, Canada

Peter Toose
Toronto
Ontario, Canada

STUDENT REPRESENTATIVES

Mahboubeh Boueshagh
Bethlehem
Pennsylvania, USA

Connor McRae-Pharo
Waterloo
Ontario, Canada

Alicia Pouw
Waterloo
Ontario, Canada

WEBMASTER

Krystopher Chutko
Saskatoon
Saskatchewan, Canada

THE PRESIDENT'S PAGE

The 80th annual Eastern Snow Conference (ESC), a joint Canada/U.S.A. organization, was held in Waterloo, Ontario 4-6 June 2024. The 80th ESC was a resounding success, held on the traditional territory of the Neutral, Anishinaabeg, and Haudenosaunee peoples. We opened the meeting with a meaningful presentation and smudge ceremony from Jean Becker in Indigenous Relations at the University of Waterloo where we had the opportunity to reflect on the lands we were on and appreciate their stewardship through the ages. All participants had the opportunity to cleanse themselves with the smudge. Snow is an integral and dynamic part of our landscapes, hydrology, ecology, and future. I, and ESC, greatly appreciate the warm welcome from our colleagues and hosts.

The meeting was successful, with 80 total registrants, 37 of whom were students. We welcomed 3 corporate sponsors, Geonor, Inc., Hoskin Scientific, and Campbell Scientific, companies that serve us as practitioners and our field in important ways. On behalf of all the participants, I would like to thank Grant Gunn, our Vice President and Program Chair, and his team of local volunteers for putting together an interesting and diverse scientific program and arranging some fun-filled social and cultural events.

Student awards are always a highlight of this conference. Congratulations to our student award winners: Lina Zschenderlein (University of Waterloo) for winning the Wiesnet Medal for best student paper, Carolyn Gorwill (Queen's University) for winning the David Miller Prize for best student poster, and Arash Rafat (Wilfred Laurier University) for the Campbell Scientific Canada Prize for showing the most innovative use of technology in the gathering of data. Congratulations to our outstanding awardees and great work to all of our student presenters who make this conference such an interesting launching point for the science of the future. We were thrilled to have Life Members Barry Goodison and Miles Ecclestone join us at the conference. In honor of our esteemed, and too early departed colleague, Josh King, we also awarded the inaugural Josh King Memorial Award for the snow and ice researcher who best exemplifies research and communication in the field. In 2024 we recognized Branden Walker (Wilfred Laurier University) who was a friend and colleague of Josh's and an outstanding exemplar of those important characteristics. Congratulations!

We were thrilled to welcome Dr. Daniel Scott, Professor and Strategic Director for Climate Change Education in Environment at the University of Waterloo, gave a lively keynote on the Tourism Side of Skiing. The keynote talk was a fascinating window into the impacts of observed and predicted snow changes on ski resort practicalities and economics. We are grateful for his contributions. Always a highlight of the week was the presentation of the coveted Sno-foo award, presented by Haejo Kim to Sam Tuttle for a variety of humor inducing incidents, including gear chewed by beavers and an unfortunate loss of a phone in a snow pit. A fitting celebration of next year's host! Congratulations.

Waterloo was a beautiful and inspiring university at which to share our collective research. Many thanks to Grant Gunn for a wonderful conference. I look forward to our next meeting hosted by Samuel Tuttle at Syracuse University in 2025. The Eastern Snow Conference is an international organization, with safe spaces for students to grow and interact with experienced snow, ice, and cold regions researchers. In 2023, we instituted a student-focused ice breaker, and we were happy to see this opportunity continue. We continue to reaffirm the commitment of the ESC to equity, diversity, and inclusion in our science and organization; we all have a role to play in making snow science a welcoming and inclusive professional home for us, the next generation, and the public.

See you next June in Syracuse, New York!

Joan M. Ramage
80th President, Eastern Snow Conference

LIFE MEMBERS OF THE EASTERN SNOW CONFERENCE

The Eastern Snow Conference gratefully recognizes individuals who have made lifelong contributions to the study of snow and for their support of this organization.

MILES ECCLESTONE

JAMES FOSTER

BARRY GOODISON

DOROTHY HALL

GERRY JONES

JOHN METCALFE

DONALD WIESNET

In memoriam

PETER ADAMS

DON DUNLAP

ART ESCHNER

AUSTIN HOGAN

HILDA SNELLING

ROBERT B. SYKES

Wiesnet Medal

The Eastern Snow Conference encourages student research through its Wiesnet Medal. This medal is presented annually to the best student paper presented at the conference.

Year	Winner	Affiliation
2024	Lina Zschenderlein	University of Waterloo
2023	Jake Ferguson	University of Waterloo
2022	Megan Verfaillie	University of New Hampshire
2021	Daniela Krampe	Alfred-Wegener-Institut
2019	Zhibang Lv	University of Saskatchewan
2018	Jiyue Zhu	University of Michigan
2017	Caroline Dolant	Université de Sherbrooke
2016	Syed Mousavi	University of Michigan
2015	Nicolas Leroux	University of Saskatchewan
2014	Justin Hartnett	Syracuse University
2013	Andreas Dietz	Earth Observation Center/DFD, Germany
2012	Elizabeth Burakowski	University of New Hampshire
2011	Kathryn Semmens	Lehigh University
2010	Simon von de Wall	University of Victoria
2009	Si Chen	Dartmouth College
2008	Chris Fuhrmann	University of North Carolina at Chapel Hill
2007	*Not awarded*	
2006	Y.C. Chung	University of Michigan
2005	M. Javan-Mashmool	Université du Québec à Chicoutimi
2004	J. Farzaneh-Dehkordi	Université du Québec à Chicoutimi
2003	Alexandre Langlois	Université de Sherbrooke
2002	Patrick Ménard	Université de Laval
2001	C. Tavakoli	Université du Québec à Chicoutimi
2000	*Not awarded*	
1999	S. Brettschnieder	Université du Québec à Chicoutimi
1998	Andrew Grundstein	University of Delaware
1997	Newell Hedstrom	University of Saskatchewan
1996	Suzanne Hartley	University of Denver
1995	Paul Wolfe	Wilfred Laurier University
1994	G.E. Mann	University of Michigan
1993	G. Devarennes	Université du Québec à Québec
1992	D.W. Cline	University of Colorado
1991	D. Samelson	Cornell University
1990	A.K. Abdel-Zaher	University of New Brunswick
1989	A. Giguere	McGill University
1988	Mauri Pelto	University of Maine
1987	Cameron Wake	Wilfred Laurier University
1986	Craig Allan	Trent University
1985	Robert Speck	Rensselaer Polytechnic Institute
1984	N.K. Kalliomaki	Laurentian University
1983	David Beresford	Trent University
1982	*Not awarded*	
1981	Jeffrey Patch	University of New Brunswick
1980	Bryan Wolfe	Trent University
1979	Margaret Leech	McGill University
1978	Michael English	Trent University
1977	Don McLaughlin & George Duggan	Rensselaer Polytechnic Institute
1976	Dwayne McMurter	Trent University
1975	Nigel Allan	Syracuse University
1974	*Not awarded*	
1973	Stan Mathewson	Trent University

David Miller Award

The David Miller Award is presented to the best student poster at the annual Conference.

Year	Winner	Affiliation
2024	Carolyn Gorwill	Queen's University
2023	Vicki Jagdeo	Lehigh University
2022	Mariah Matias	Lehigh University
2021	Madison Woodley	Syracuse University
2019	Gaohong Yin	University of Maryland
2018	Justin Pflug	University of Washington
2017	Elizabeth Ryan	University of Maryland
2016	Oliver Wigmore	The Ohio State University
2015	Charles Papasodoro	Université de Sherbrooke
2014	Laura Thomson	University of Ottawa
2013	Nastaran Saberi	University of Waterloo
2012	Johnathan Sugg	Appalachian State University
2011	*Not awarded*	
2010	Nicolas Svacina	University of Waterloo
2009	Iliyana Dobreva	Texas A&M University
2008	Christopher Karmosky	Pennsylvania State University

Campbell Scientific Canada Prize

The Campbell Scientific Canada prize is awarded to student research showing the most innovative use of technology in the gathering of data.

Year	Winner	Affiliation
2024	Arash Rafat	Wilfred Laurier University
2023	*Not awarded*	
2022	Alicia Pouw	Wilfred Laurier University
2021	Fraser King	University of Waterloo
2019	Alex Mavrovic	Université de Sherbrooke
2018	*Not awarded*	
2017	Abigail Dalton	University of Ottawa
2016	*Not awarded*	
2015	*Not awarded*	
2014	Aaron Campbell	University of Waterloo
2013	*Not awarded*	
2012	Andrew Kasurak	University of Waterloo
2011	Mathieu Beaulieu	University of British Columbia
2010	*Not awarded*	
2009	*Not awarded*	
2008	Nicholas Kinar	University of Saskatchewan
2007	Gro Lilbaek	University of Saskatchewan
2006	Gernot Koboltschnig	University of Natural Resources and Applied Life Sciences, Vienna
2005	Warren Helgason	University of Saskatchewan

Josh King Memorial Award

The Josh King Memorial Award is presented to a snow and ice researcher that best exemplifies research and communication in the field.

Year	Winner	Affiliation
2024	Branden Walker	Wilfred Laurier University

Oral Session #1

Session Chairs: Benoit Montpetit & Madison Woodley

Impact of Seasonal Snow Accumulation on Lake Ice Development in Canada's Sub-Arctic during 2023-24 Winter Season

ALICIA POUW[1,2] AND HOMA KHEYROLLAH POUR[1,2]

ABSTRACT

Snow accumulation on lake ice significantly impacts ice growth and melt timing due to its insulative and reflective properties. Therefore, variations in lake ice thickness are expected due to heterogeneous accumulation of snow, caused by factors such as wind redistribution, snowpack metamorphism, and un-uniform snow ice formation. Although current lake ice models effectively simulate ice thickness, they demonstrate high sensitivity to variations in snow depth and density. Given the continuous processes affecting lake ice development, there is a critical need for snow and ice observations with high spatial and temporal resolution to improve lake ice model parameterizations. This study is supported by field campaigns conducted monthly from November 2023 to April 2024, collecting in-situ snow and ice surveys across 5 sub-Arctic lakes surrounding Yellowknife, NT. These surveys provide essential data for examining how seasonal snowpacks affect lake ice development. Preliminary results suggest lakes that freeze earlier and are not frequently travelled, experience an increase in snow ice formation consistently throughout the winter season (25 – 45% vs. 7 – 25% of total ice thickness). By March, however, the total ice thickness becomes comparable across the lakes studied, with more congelation ice growth occurring on lakes that experienced less snowfall from a later freeze up (~15 % more congelation ice growth). Additionally, lakes that experienced regular snowmobiling actively throughout the season measured higher snow densities and minimal snow ice formation. The findings from this ongoing research can lead to enhance understanding of snow-lake ice interactions, which is essential for lake ice model improvements and supporting the safety and well-being of northern communities within a changing climate.

[1] Remote Sensing of Environmental Change Research Group, Department of Environmental Science, Wilfrid Laurier University, Waterloo, ON, Canada

[2] Cold Regions Research Centre, Wilfrid Laurier University, Waterloo, ON, Canada

Corresponding author: apouw@wlu.ca

Merging Land Surface Models and Snow Pit Data for Improved Snow Surface Density Measurements for Wildlife Tracking Purposes

JACK DECHOW[1], MICHAEL DURAND[1], JESSICA LUNDQUIST[2], LAURA PRUGH[2], BEN SULLENDER[2], AND CALLUM CUNNINGHAM[3]

ABSTRACT

Wildlife movement patterns in Snow covered enviroments are directly impacted by the conditions of the snowpack. FOr each animal species There is a surface density threshold that determines whether the snowpack can support an ANIMAL'S weight. Thus, the spatial variability of snow surface density is a control mechanism on the pathing of wildlife. For our study area in central Washington STATE, we present a parameterized non-linear state model to merge snow pit data from an extensive field campaign with high resolution (30m) estimates of snowpack characteristics from the NoahMP land surface model (LSM) ran in the nasa land interface system (LIS) framework. Initial snow surface and vegetation temperatures from the LSM are assimilated with 1km MODIS Land Surface Temperature data products to generate a better estimate of snow surface and vegetation temperature. The state model is parameterized with SWE, snowfall, and surface density from the LSM, the assimilated snow temperatures, and surface density from the field campaign. The parameterized state model is then used to generate new surface density estimates gridded at 30m. For water year 2022 we present error statistics showing the improved accuracy of the updated surface density results when compared with the field dataset. FURTHERMORE, we generate spatial maps highlighting areas where LSM estimates of surface density missed the support threshold while the updated surface density estimates correctly identify these areas.

[1] The Ohio State University, Columbus, OH, USA
[2] University of Washington, Seattle, WA, USA
[3] University of Tasmania, Hobart, Tasmania, Australia
Corresponding author: dechow.5@osu.edu

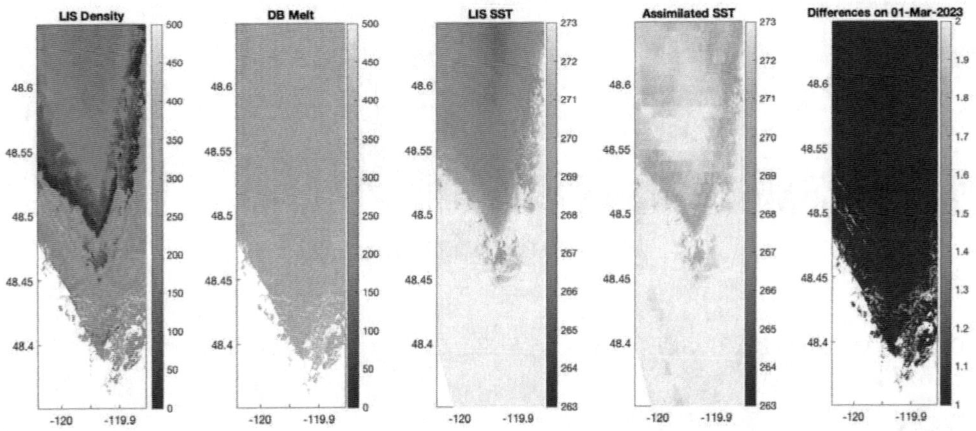

Figure 1. From left to right: Surface Density Estimate from LIS, Parameterized Surface Density Estimate, Snow Surface Temperature from LIS, Assimilated Snow Surface Temperature, Density Threshold differences (1 = LIS below threshold and Parameterized above threshold, 2 = Both above threshold) for March 1st 2022.

Coupling of MDSS and AVL to Achieve Dynamic Maintenance Recommendations

DIAR HASSAN[1], KARL SCHENABEL[1], SHAHIN ABJI[2], AND ELIZABETH LODER[3]

ABSTRACT

The Maintenance Decision Support System (MDSS) is an ideal tool for road authorities to guide their winter maintenance operations and formulate their daily logistic planning. Leveraging each agency's unique Rules of Practice with forecasted road conditions, MDSS acts as a virtual handbook, suggesting the optimal maintenance activities that will meet required Service Levels. The one caveat is that MDSS relies on forecasted treatments, which become outdated as conditions and constraints evolve in real time and other application times/rates are utilized. Automatic Vehicle Location (AVL) data provides geospatial insights to each vehicle's activity, such as the spatial location of the truck, blade(s) status, spreader status, chemical type, and chemical rates. WSP has been working with Alberta Transportation (AT) on a pilot project to couple the real-time AVL data with the MDSS modules spatiotemporally. The result is powerful, dynamic recommendations that respond as new maintenance activities are undertaken and update all downstream recommendations to reflect the changing conditions. Pilot results will be presented alongside insights into the future of this intelligent technology.

[1] WSP Canadan Inc., Ottawa, ON, Canada
[2] Alberta Transportation, Edmonton, AB, Canada
[3] WSP Canadan Inc., St. John's, NL, Canada
Corresponding author: diar.hassan@wsp.com

How Snow has Changed Over 30 Years at Trail Valley Creek, and what this means For Streamflow And Active Layer Thaw

BRAMPTON DAKIN[1], PHILIP MARSH[1], DAVID RUDOLPH[2], ROBIN THORNE[1], AND BRANDEN WALKER[1]

ABSTRACT

Earth's dramatically warming climate is impacting many aspects of the Arctic environment at unprecedented rates. We have started to see changes impacting active layer depths and permafrost, shrubs have started to expand north. These are interacting to produce complex changes to all aspects of Arctic hydrology. Unfortunately, the small number of modelling stations across the Arctic make it difficult to understand the overall impact of these changes and it is still unknown whether these impacts will result in the wetting or drying of the Arctic. This study aims to analyze and model the 30-year record of meteorological, streamflow, and snow water equivalent (SWE) data at the Trail Valley Creek research watershed (TVC) in the NWT. This study will look at how snow over the past 30 years has changed, and using the Arctic hydrological model, GEOtop, we will analyze how this has affected active layer depths, evapotranspiration, and streamflow across the entirety of the 55km^2 TVC watershed. This includes across different landcover types such as lakes, snow drifts, tundra, and taller / shorter shrub patches. Using these results, we will then evaluate whether the catchment is becoming more wet or dry as a whole, and if different landcover types are changing with this.

[1] Cold Regions Research Centre, Wilfrid Laurier University, Waterloo, ON, Canada
[2] Cold Regions Research Centre, University of Waterloo, Waterloo, ON, Canada
Corresponding author: bdakin@wlu.ca

Causes of Winter and Spring Peak Streamflows in the Contiguous United States

MEGAN VERFAILLIE[1] AND JENNIFER M. JACOBS[2]

ABSTRACT

Climate change is impacting many wintertime and cold region flood drivers, such as air temperature, precipitation patterns, snowmelt timing, and ice formation and breakup. Combined with a growing concern for increased frequency and severity of precipitation events, there is a need for better understanding of when and why these events occur in cold regions. For many flood events, relevant flood conditions (e.g., soil saturation, snow water equivalent) go undocumented, and flood risk assessments that only account for individual drivers of flooding may lead to an underestimation of risk. This is especially true in winter and spring where the presence of snow and ice, frozen ground, freezing temperatures, and mixed precipitation make conditions more complex. This study used remotely sensed and modeled data to perform flood attribution for the largest winter and spring peak streamflow events at 71 USGS gaged drainage basins between 2003 – 2022. Attribution was determined using NASA NLDAS-2 modeled precipitation, air and soil temperature, and soil moisture, NOAA NWS NOHRSC SNODAS modeled snow water equivalent (SWE), and gap-filled MODIS 1-km snow cover observations. Results were validated using Dartmouth Flood Observatory (DFO) records. Unique events with overlapping records from DFO (19/71 basins) showed good agreement with validation data (14/19 matching causes). The most common flood driver was rainfall. When snow was present, the most common events had all flood drivers (i.e., rainfall, frozen ground, saturated soils, snow cover, SWE) present as compared to only rain and snow. These results provide the foundation for an investigation of current observational data gaps for differentiating between minor and extreme winter and spring floods.

[1] Department of Civil and Environmental Engineering, University of New Hampshire, Durham, NH, USA

[2] Earth Systems Research Center, Institute for the Study of Earth, Oceans, and Space, University of New Hampshire, Durham, NH, USA

Corresponding author: megan.verfaillie@unh.edu

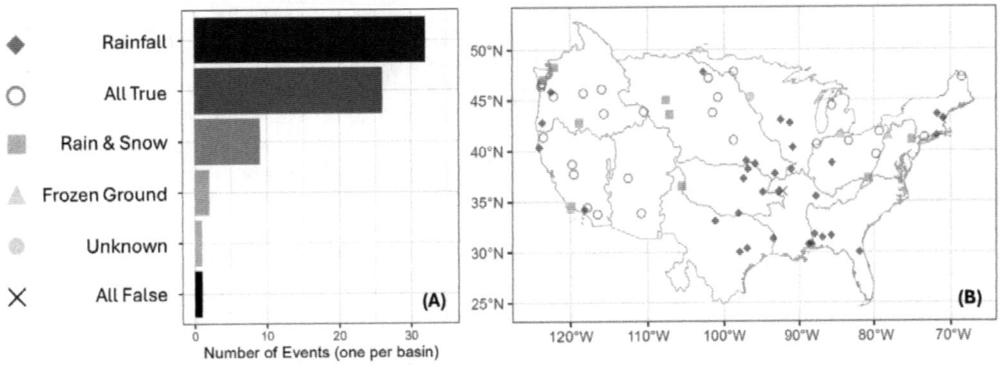

Figure 1. Flood attribution results for 71 unique, winter and spring peak streamflow events. "All True" indicates that all flood drivers (i.e., rainfall, frozen ground, saturated soils, snow cover, SWE) were present leading up to the event. **(A)** Number of peak streamflow events in each attribution category. **(B)** Gage locations indicated by the attribution result for their winter or spring peak flow event. Gray lines outline the National Weather Service River Forecast Center boundaries.

Poster Session #1

Session Chair: Eli Deeb

Advancing Snow Water Equivalent (SWE) Estimation: Sensitivity Testing of Dual Ku-Band Radar System

NICOLAS ALLET[1,2], ALEXANDRE LANGLOIS[1,2], DANIEL KRAMER[1,2], BENOIT MONTPETIT[1,3], AND JULIEN MELOCHE[3]

ABSTRACT

As part of the University Consortium for the Canadian Terrestrial Snow Mass Mission (TSMM), this study contributes to the mission's overarching goal of advancing our understanding of terrestrial snow mass dynamics in the context of climate change. Focused on evaluating the sensitivity of radar measurements to snow properties, particularly snow water equivalent (SWE), the Université de Sherbrooke have recently acquired two Frequency Modulated Continuous Wave (FMCW) radars, emulating the proposed dual-Ku antennas for TSMM. These FMCWs were deployed during the winter 2023-24 campaigns, and particularly in Cambridge Bay, Nunavut, Canada, in April 2024.

This field campaign presents a unique opportunity to characterize the Ku-band radar measurements to Arctic snow properties. These measurements will also provide valuable data to improve microwave radiative transfer modeling, using the Snow Microwave Radiative Transfer (SMRT) model.

Prior to the field campaign, sensitivity tests, using SMRT and field measurements from previous field campaigns in Cambridge Bay, were conducted to predict the behavior of the radar signal to snow properties. This presentation will present sensitivity analysis and the acquired field measurements. This will provide insights on the effectiveness of radar-based approaches for SWE estimation and the limitations and improvements needed to snow microwave radiative transfer modeling. Ultimately, our findings will contribute valuable data and methodologies to the TSMM mission, enhancing its capacity to monitor terrestrial snow mass and advance our understanding of Earth's water cycle dynamics in the face of climate change.

[1] Centre d'Applications et de Recherches en Télédétection (CARTEL), Université de Sherbrooke, QC, Canada

[2] Groupe de Recherche Interdisciplinaire sur les Milieux Polaires (GRIMP)

[3] Climate Change Division, Environment and Climate Change Canada (ECCC)

Corresponding author: Nicolas.allet@usherbrooke.ca

Spatial Heterogeneity in Snowpack Response to Rain-on-Snow Events in a Boreal Forest of Eastern Canada

DMITRY PERSHIN[1,2], DANIEL F. NADEAU[1,2], MICHEL BARAËR[3], BENJAMIN BOUCHARD[1,2], AND PIERRE-ERIK ISABELLE[1,2]

ABSTRACT

Winter rain-on-snow (ROS) events can trigger significant flooding in snow-dominated catchments, but the occurrence of ROS floods varies depending on factors such as rainfall intensity and the contribution of snowmelt to runoff across the catchment. In boreal forests, land cover is often patchy due to logging and fires, affecting snowpack characteristics. However, the snowpack response to ROS in such environments remains unclear. This study examines the influence of storm characteristics and initial snowpack conditions on rainwater behaviour, snowpack properties, and snow water equivalent (SWE) evolution during ROS events on December 9 and 17, 2023, in the Montmorency Forest (47° N, 71° W), Quebec, Canada. The snowpack ROS response is compared in a 400 m-wide clearcut and a 10 m-wide forest gap within a juvenile balsam fir stand with a 9 m canopy height. We used automated observations of snow depth, temperature, SWE, and snow pit data. The first event (ROS1) brought 14 mm of rainfall over nearly 17 hours, while the second event (ROS2) was much larger, with 105 mm of rainfall within 40 hours. During ROS1, both sites gained SWE, suggesting that a significant portion of the rainwater refroze. ROS2 resulted in the clearcut site losing 41 mm of SWE, while the gap site lost only 13 mm. Water passed through the snowpack faster at the clearcut than at the gap. Rainwater reached the snow based before the snowpacks became isothermal at both sites, indicating preferential flow and rainwater bypassing energy and mass exchange with the snowpack. Both events brought the snowpacks to an isothermal state, but the severity of the streamflow response was primarily driven by rainfall rates.

[1] Department of Civil and Water Engineering, Université Laval, Quebec City, QC, Canada
[2] CentrEau – Water Research Centre, Université Laval, Quebec City, QC, Canada
[3] Department of Construction Engineering, École de technologie supérieure, Montréal, QC, Canada
Corresponding author: dmitry.pershin.1@ulaval.ca

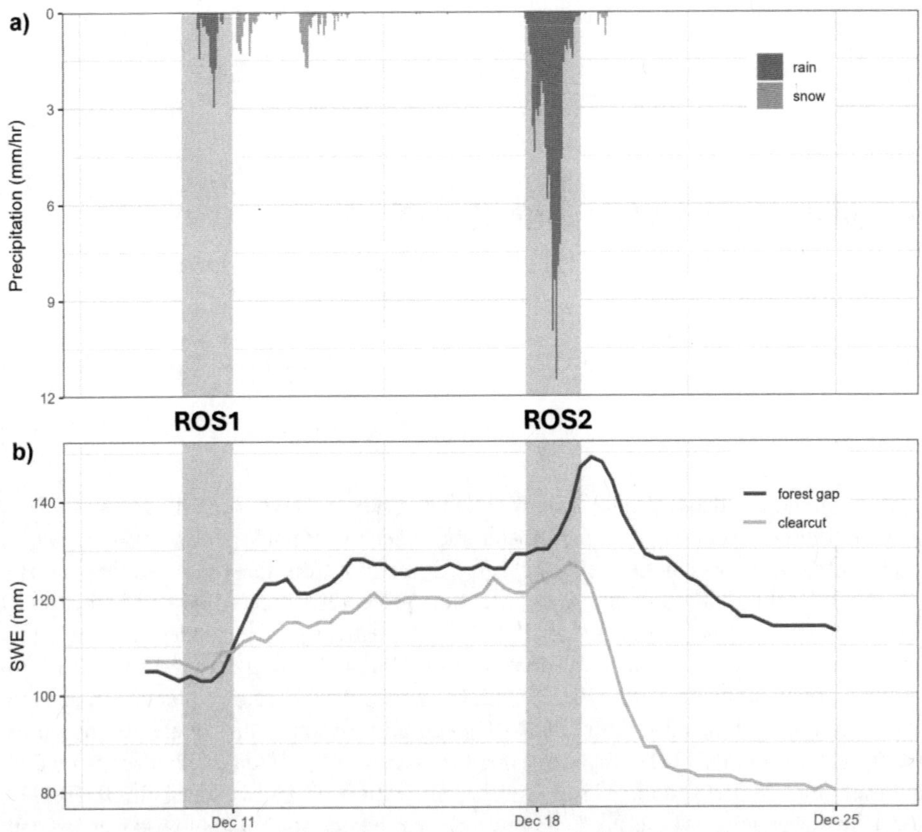

Figure 1. Rainfall and snowfall rates (a) and SWE change (b) at clearcut and forest gap sites in the Montmorency Forest during ROS1 and ROS2. Precipitation and phase data were obtained from the NEIGE site in the Montmorency Forest (https://www.cef-cfr.ca/pmwiki.php?n=Membres.SylvainJutrasNEIGE), while SWE measurements were performed using the CS725 SWE sensor (Campbell Scientific).

How well do the NASA Standard MODIS and VIIRS Snow-Cover Products compare for Creation of a Climate-Data Record?

DOROTHY K. HALL[1,2], GEORGE A. RIGGS[3], AND NICOLO E. DIGIROLAMO[3]

ABSTRACT

A nearly-continuous daily, global Environmental Science Data Record (ESDR) of NASA Standard MODerate-resolution Imaging Spectroradiometer (MODIS) snow-cover extent (SCE) data products has been available since 24 February 2000, first from the Terra satellite, and starting in 2002, also from the MODIS on the Aqua satellite. The MODIS record will end when the Terra and Aqua satellites are decommissioned in 2026 or 2027. However, the 'moderate resolution' SCE product record will continue with the Suomi-National Polar-orbiting Partnership (NPP) Visible Infrared Imaging Radiometer Suite (VIIRS) NASA Standard SCE data products, thus enabling a 30+ year record of moderate resolution SCE which could potentially be elevated to the status of a Climate Data Record (CDR). Here, we focus on the NASA Standard daily cloud-gap-filled (CGF) 500-m resolution MODIS and 375-m resolution VIIRS SCE products, MOD10A1F and VNP10A1F, respectively. These products have been used for studying snow-cover persistence, annual and interannual variability in SCE, and regional and hemispheric trends in SCE. The 11+ year overlap period of the MOD10A1F and VNP10A1F products provides a valuable opportunity to assess agreement between the data products. Our results show a high degree of agreement between the two products in the Great Basin of the western United States. The number of days (#days) of snow cover per month, and monthly snow-cover extent (SCE), calculated from the MOD10A1F and VNPA1F products, is very similar for the 11-year study period from 2012-2023, showing agreement with a Pearson's correlation, R = 0.998, for #days of snow cover and R = 0.991, for SCE. In general, the VNP10A1F maps slightly more days of snow cover and slightly greater SCE than does MOD10A1F. By the measures described above (#days of snow cover and SCE), the NASA Standard SCE data products display good agreement and high continuity over the period of sensor overlap beginning in 2012.

[1] Earth System Science Interdisciplinary Center/University of Maryland, College Park, MD, USA
[2] Cryospheric Sciences Lab / NASA / Goddard Space Flight Center, Greenbelt, MD, USA
[3] SSAI, Lanham, MD, USA
Corresponding author: dkhall1@umd.edu

Machine Learning and Snow Temperature: Snow Depth Estimation in a Forested Watershed in the Adirondack Mountains, NY

MADISON WOODLEY[1], MICHAEL MAHONEY[2], JAMES S. MILLS[2], COLIN BEIER[2], AND SAMUEL TUTTLE[1]

ABSTRACT

The northeastern United States has recently experienced warming winters, leading to shorter snow seasons punctuated by midwinter melt events. Understanding the spatiotemporal distribution of a snowpack as it naturally varies within a watershed can enhance hydrologic prediction and reveal ecological implications. We use low-cost temperature sensor profiles paired with cameras to observe snow energy state and estimate snow depth in the Arbutus Lake watershed, located in the central Adirondack Mountains of New York, USA. Paired temperature sensor profiles were installed at 24 locations in a grid pattern within the upper portion of the watershed in 2019 to monitor snow and soil microclimate. Each profile consists of low-cost iButton temperature sensors at heights of -20 cm to 100 cm above the ground surface enclosed in PVC piping at 20 cm spacing. Several of the temperature profiles were paired with trail cameras and snow stakes starting in winter 2021 - 2022 for daily snow depth (SD) estimation. We use these paired photographs and temperature profiles to train random forest (RF) machine learning models to estimate SD solely from snow temperature profiles, providing distributed SD estimates across the watershed from 2019 to present. Our initial RF model fits show that temperature profiles with 20 cm sensor spacing can be used to infer SD with uncertainty of 5-10 cm. We observed minimal differences in temperature data between temperature profiles shielded within PVC tubes versus temperature profiles exposed directly to the snowpack. Future work will evaluate canopy gap fraction and solar radiation indices as additional independent variables in the RF models of SD.

[1] Department of Earth & Environmental Sciences, Syracuse University, Syracuse, NY, USA
[2] SUNY College of Environmental Science and Forestry, Syracuse, NY, USA
Corresponding author: mwoodley@syr.edu

Evaluating a Hierarchy of Bias Correction Methods for Reanalysis SWE Estimates in Northern Canada

NEHA KANDA[1] AND CHRISTOPHER G. FLETCHER[1]

ABSTRACT

Precise estimates of snow water equivalent (SWE) are crucial for informed decision-making in regions like Northern Canada, where snow cover significantly contributes to springtime discharge. However, the sparse nature of the existing SWE monitoring network poses a challenge to comprehensively understanding the SWE distribution and variability. Reanalysis products like ERA5-Land provide long-term continuous SWE estimates, but our evaluation identified a negative bias (-61 mm) in the estimated SWE and maximum underestimation was observed at high elevation (>1500 m) areas. To correct these biases, we applied four correction methods: mean bias subtraction (MBS), simple linear regression (SLR), multiple linear regression (MLR), and random forest (RF). RF exhibited the highest performance, reducing the root mean square error (RMSE) by 78% and minimizing the annual mean bias from 61.2 mm to 0.01 mm. RF showed limited spatial transferability to regions characterized by very low or very high SWE values. However, it was able to capture the typical SWE values at most ecoregions. The bias-corrected SWE dataset will be used to validate and improve the estimation of snowmelt runoff, and to gain an improved understanding of the impact of SWE biases on snowmelt and subsequent spring soil moisture.

[1] Department of Geography & Environmental Management, University of Waterloo, Waterloo, ON, Canada
Corresponding author: nkanda@uwaterloo.ca

Snow Depth from Spaceborne Lidar: Insights from ICESat-2 over the Boreal Forests and Tundra of Alaska

ZACHARY FAIR[1], CARRIE VUYOVICH[1], AND TOM NEUMANN[1]

ABSTRACT

Lidar is an effective tool to measure snow depth over key watersheds across the United States. Lidar-derived snow depth observations from airborne platforms have demonstrated centimeter-level accuracy at high spatial resolution, and in recent years there has been an emerging interest in using spaceborne lidar to estimate snow depth. Results from spaceborne lidar altimeters such as NASA's Ice, Clouds, and Land Elevation Satellite-2 (ICESat-2) can provide accurate snow depth retrievals over mid-latitude watersheds, though assessments over boreal forest and tundra environments have been limited. Also, further research on the accuracy, coverage, and operational potential is needed.

In this presentation, we outline ICESat-2 snow depths derived over the boreal forest and tundra of Alaska. We focus on two field sites selected by the SnowEx 2022/2023 campaign: Creamer's Field near Fairbanks, AK and Upper Kuparuk/Toolik on the Arctic Coastal Plain. We assess three ICESat-2 products (ATL06, ATL08, SlideRule) relative to airborne lidar data collected by the University of Alaska, Fairbanks (UAF). Generally, we observed the best performance in ICESat-2 over the tundra region (Toolik). Agreement between ICESat-2 and UAF was weaker over the boreal forest (Creamer's Field), with dense vegetation thought to be responsible for observed errors. We use these results to summarize best practices for spaceborne lidar snow depth retrievals, as well as ongoing challenges (sloped terrain, vegetation, etc.). We conclude with a discussion on the feasibility of using spaceborne lidar to fulfill Decadal Survey requirements, with suggestions for future studies and requirements for future mission concepts.

[1] NASA Goddard Space Flight Center, Greenbelt, MD, USA
Corresponding author: zachary.fair@nasa.gov

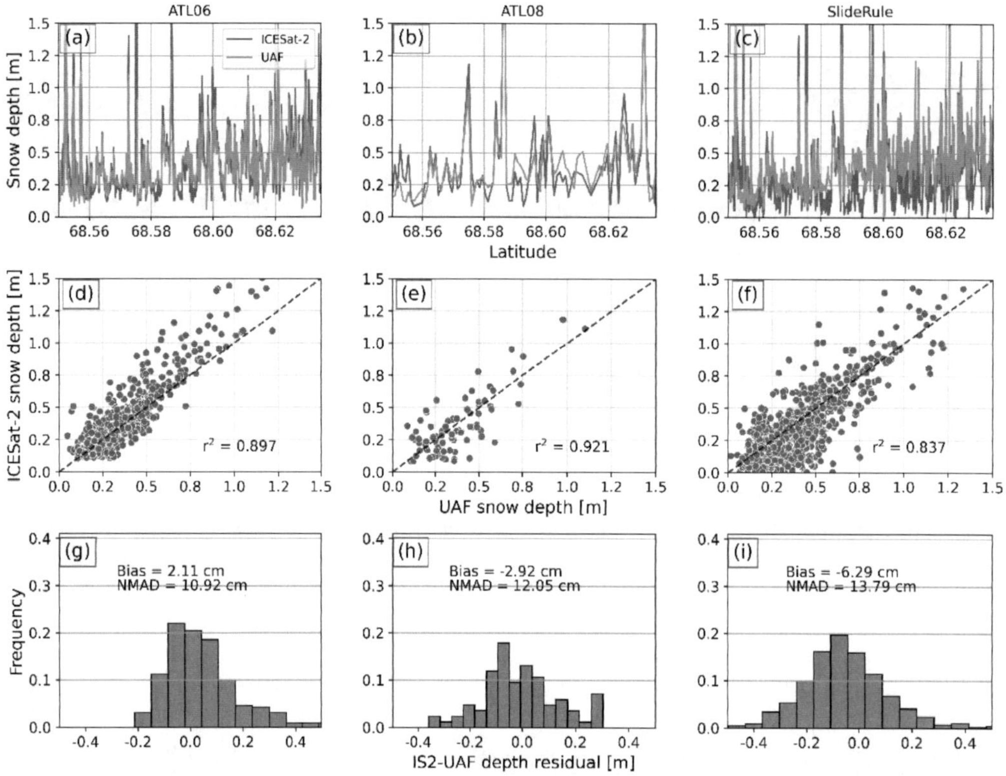

Figure 1. Snow depth results over Upper Kuparuk/Toolik, using ICESat-2 track 152 from March 31, 2023 and airborne lidar from UAF. (a-c) Along-track snow depth results from ICESat-2 (blue) and UAF lidar (orange) from the ATL06 (a), ATL08 (b), and SlideRule (c) products. (d-f) Scatter plots of the two datasets for the three ICESat-2 products, with r^2 values also given. (g-i) Histograms of ICESat-2 depth residuals, with median bias ("Bias") and normalized median absolute deviation ("NMAD") given.

New VIIRS 375 m Global Snow Cover Product, and Update on MODIS Passivation and Product Production

GEORGE A. RIGGS[1], VIRGINIA L. KALB[2], AND DOROTHY K. HALL[3]

ABSTRACT

A new Visible Infrared Imaging Radiometer (VIIRS) daily 375 m global snow cover product has been developed for Suomi-National Polar-orbiting Partnership (SNPP) and NOAA-20. This daily global product is produced by reprojecting the observations in the VIIRS daily tiled product, VNP10A1, for a day to a global geographic projection to create the new VNP10D1 at 375 m resolution. The VNP10D1 algorithm does not bin the VNP10A1 observations to 5 km resolution as is done for the daily Climate Modeling Grid (CMG) VNP10C1 data product. Unlike the VNP10C1 data product which has separate data layers for the snow cover and cloud cover the VNP10D1 has only a snow cover data layer that includes cloud observations. The VIIRS 375 m global snow cover product is presented for both the SNPP and NOAA-20.

An update on NASA plans for passivation of Terra and Aqua MODerate-resolution Imaging Spectroradiometer (MODIS), and data product collection processing is presented. Both Terra and Aqua are in orbital drift, Phase F of operation, until sometime in 2026. The timeline for passivation is updated from the timeline presented at the ESC 2023 to aid users in planning to transition from MODIS to VIIRS data products.

---

[1] SSAI, Lanham, MD, USA
[2] NASA GSFC Code 619, Greenbelt, MD, USA
[3] Earth System Science Interdisciplinary Center/University of Maryland, College Park, MD, USA
Corresponding author: george.riggs@ssaihq.com

Observing and Predicting the Snow Microclimates of an Ephemeral Snowpack in Southeastern New Hampshire

JEREMY JOHNSTON[1], JENNIFER M. JACOBS[1,2], ADAM HUNSAKER[1], MAHSA MORADI[2], TIM HOHENEDER[1,3], KARA WITTMANN[2], KATIE BATEMAN[2], SYDNEY VANASSE[4], AND AIMEE BOUSQUET[5]

ABSTRACT

Snow properties, soil, and surface temperatures can vary widely, especially in areas with mixed land cover and terrain. Such variations, termed 'microclimates', frequently occur across short distances (< 1 km) due largely to local differences in topography, soils, and vegetative cover. Over these small domains, traditional sensing approaches rely on individual meteorological stations to validate land surface models and satellite observations. Restrictive cost or effort usually limits the implementation of dense *in situ* networks at such scales. To address these challenges and to effectively monitor microclimates, we developed a low-cost network of *in situ* sensors to measure soil temperature profiles, snow depth, humidity, and air temperature within a mixed forest and open field study area (~0.5 km^2) in southern New Hampshire. Using terrain and forest structural variables derived from Uncrewed Aerial System (UAS) lidar observations, *k*-means clustering was used to place these *in situ* stations in the most numerically unique areas. To assess the representativeness of an individual station, observations from our network over the winter of 2023-24 are first compared against those from a co-located United States Climate Reference Network (USCRN) station. Subsequently, we rely on snow depth and persistence observations collected via UAS and regular observations from ground sampling crews to assess how well clustering techniques predict local areas of similar snow accumulation and melt properties within an ephemeral (i.e., warm/shallow) snowpack. Results show notable differences in soil temperature distributions, snow depth, and snow persistence, depending on the land cover types. This effort provides a case study on the drivers of snow microclimates which can inform improved model representations of snow and soil energy balances while offering a strategy to predict local-scale snow climates in a computationally efficient manner.

[1] Earth Systems Research Center, Institute for the Study of Earth, Oceans, and Space, University of New Hampshire, Durham, NH, USA
[2] Department of Civil and Environmental Engineering, University of New Hampshire, Durham, NH, USA
[3] University of New Hampshire Graduate School, University of New Hampshire, Durham, NH, USA
[4] Department of Earth Sciences, University of New Hampshire, Durham, NH, USA
[5] College of Life Sciences and Agriculture, University of New Hampshire, Durham, NH, USA
Corresponding author: jeremy.johnston@unh.edu

INTRODUCTION

Even in areas with similar meteorologic forcings (e.g., air temperature, humidity, precipitation, incoming shortwave and longwave radiation), surface temperatures and snow properties vary widely due to hillslope scale or smaller (<100 m) vegetation and terrain influences. Nearby areas (< 1 km) with persistent differences in snow properties are termed 'snow microclimates.' At these scales, shading, precipitation interception by the canopy, and wind sheltering cause variations in local climate conditions which affect snowpack evolution. However, variability at these scales is difficult to capture with traditional ground-based or remote-sensing approaches. Manual efforts to sample snow variability across such regimes are common (Clark et al., 2011), but they can be time and labor-intensive and do not provide continuous observations. Resource limitations (e.g., cost) usually prevent the installation of continuous ground-monitoring networks capable of capturing hillslope scale variability in meteorologic and snow conditions. Thus, large regional networks such as the United States Climate Reference Network (USCRN) or the Snow Telemetry Network (SNOTEL) are the standard approach for collecting continuous ground observations. Such networks rarely include stations located within 5 km of one another.

Most ephemeral snow areas have mean air temperatures near 0 °C. As a result, these areas are characterized by shallow snow with frequent snowfall and snowmelt events (Sturm and Liston 2021; Johnston et al., 2023). These characteristics also result in high spatiotemporal snowpack variability. Achieving the high spatial and temporal resolution required to observe small variations in hillslope-scale snowpack evolution in ephemeral snow climates thus requires a combination of multiple data sources. *In situ* automated sampling networks offer temporally continuous observations and Uncrewed Aerial System (UAS) provide high-spatial resolution observations. UAS lidar provides site-wide observations of vegetation, terrain, and snow characteristics (e.g., snow depth, snow-covered area). In contrast, automated stations continuously observe near-surface meteorology and snow conditions.

The objectives of this study were (1) to assess the hillslope-scale variability in air and soil temperatures, (2) to identify the hillslope-scale controls on variability in snow depth and persistence, and (3) to assess the applicability of multivariate clustering for optimizing meteorologic station placement within an ephemeral snow regime. To address these objectives, the following analysis compares snow depth and persistence observations from *in situ* stations and UAS to various vegetation and terrain variables.

DATA AND METHODOLOGY

The study area was Kingman Farm, a 0.5 km^2 site in Madbury, New Hampshire (43.172° N, 70.928° W) maintained by the University of New Hampshire. It consists of a diverse mixture of managed forests and fields. The forests have a range of height, density, age, and type (e.g., evergreen, deciduous, mixed species). Investigations in fall 2023 from site visits, UAS-optical imagery, and *k*-means clustering of UAS-lidar observations informed the placement of a dense network of automated meteorological stations for winter 2023-2024 (Figure 1). Clustering used a subset of the variables shown in Table 1 to identify seven intensive sampling sites. Seven low-cost Arduino-based meteorologic stations were installed in these locations. The stations included air temperature, relative humidity, soil temperature (0, 5, 15 cm), ultrasonic snow depth sensors (HC-SR04), and a Wingscapes timelapse camera. Additional stations (x2, Campbell Scientific instrumentation) included snow and soil temperature profiles, air temperature, snow depth (Judd Communications), and Wingscapes timelapse cameras. One co-located station was grass-covered, and one was stripped to the bare soil to quantify the influence of vegetative cover on the near-surface thermal regime. Timelapse imagery was collected at 30-minute intervals, and other sensors collected observations at a 5-minute sampling frequency between January 5 and April 15, 2024. The network was co-located with a long-term USCRN station (NH Durham 2 N, *https://www.ncei.noaa.gov/access/crn/*) and was monitored regularly throughout the winter to ensure data quality. Snow observations and optical imagery were not collected at the CRN station.

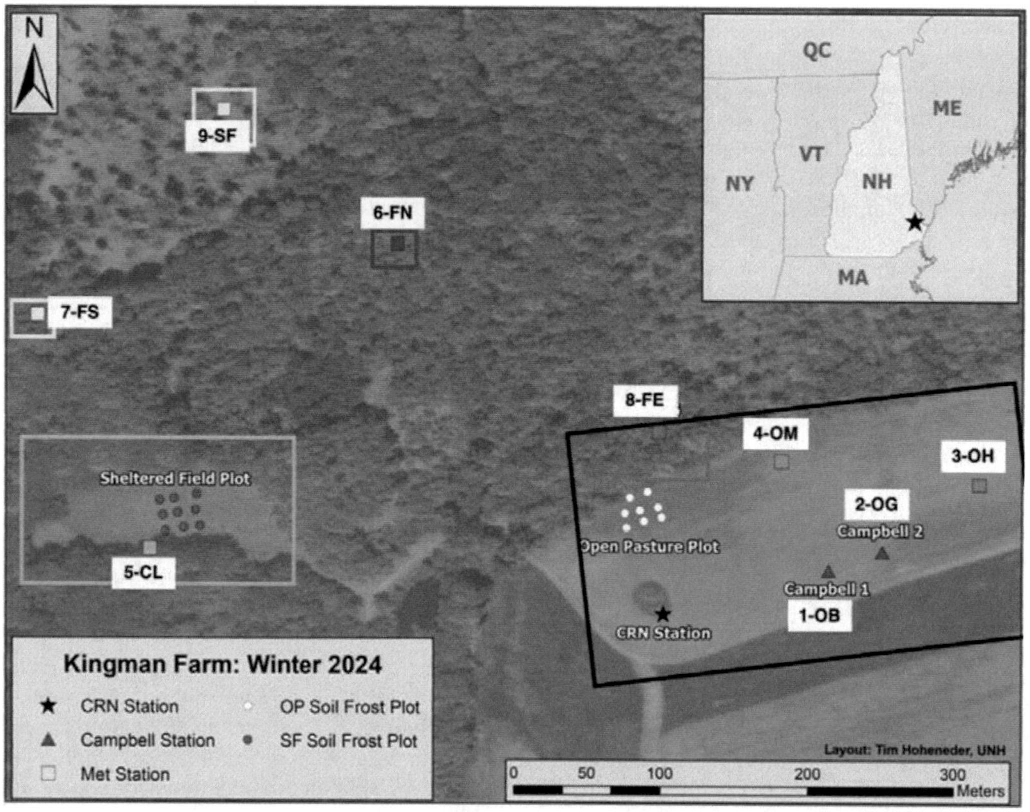

Figure 1. The Kingman Farm study area with automated *in situ* sampling stations indicated. Open (O) field stations are bounded in black, closed field (5-CL) in dark yellow, sparse forest (9-SF) light yellow, north-facing forest (6-FN) in dark green, south-facing forest (7-FS) in light green, and forest edge (8-FE) in green. For open field stations, 1-OB indicates 'open bare soil,' 2-OG indicates 'open grass,' 3-OH indicates 'open high-exposure,' and 4-OM indicates 'open moderate-exposure.' Squares indicate Arduino-based samplers, triangles Campbell scientific equipment and loggers, and the black start indicates the Climate Reference Network (CRN) Kingman Farm station.

UAS lidar observations were collected using a GreenValley International LiAir V70 mounted on a DJI M300. The lidar point clouds produced descriptive terrain and vegetation metrics across Kingman Farm. They were also used to estimate snow depths on two dates (January 8, 2024, and January 30, 2024). Bias-corrected snow depths had errors of approximately 3 cm in open areas and 5 cm in forested areas. Optical imagery was also collected using a Phantom 4 RTK. Orthophotos were produced from the imagery covering Kingman Farm on nine dates with partial snow cover conditions. Supervised image thresholding was used to delineate snow-on and snow-free areas from the orthophotos. A snow persistence map was produced as the proportion of all nine flights with observed snow cover. All UAS flight dates are indicated in Figure 2.

Rigorous QA/QC was performed on each collected temperature and snow depth dataset (stations 1-9). Errant values were removed, and temperature sensor stability was assessed through visual inspection of all temperature time series. Temperature sensors were calibrated following field removal using a two-point calibration (ice bath and boiling water), and sensor error was reduced to within 0.25 °C. Snow depths were estimated approximately twice daily by trained observers using a stake marked at 5 cm intervals within the images. A higher frequency of estimates was recorded during periods of rapid change (e.g., melt, accumulation). The observers also estimated the snow cover percentage to be the proportion of snow on the ground within the image scene. All estimates were verified by comparing between multiple independent observers. Data from all stations was matched to the same period in which all sites had observations (January 5 – April 15, 2024). The mean, standard deviation, and skew were used to compare the temperature distributions. The two-sample Kolmogorov-Smirnov (Massey Jr., 1951) test statistic was used to compare distribution similarity between stations 1-9 and the CRN station. The duration of frozen soil, defined as the number of days in which 5 cm depth soil temperatures were below 0 °C, was also compared across all stations. Using only stations with snow depth and snow cover data, Spearman's rank correlation coefficient (Kendall, 1948) was used to compare numerous snow metrics to local vegetation and terrain metrics. Station snow metrics included:

I) snow cover duration (SCD): sum of days during the study period with snow cover;
II) core snow season (CSS): the length in days of the longest unbroken period of snow cover;
III) mean snow depth (MSD): the average depth of snow cover considering only periods with snow;
IV) mean maximum snow depth (MMSD): the average depth of snow considering only the maximum values following each of the seven observed storm events.

Installed stations spanned a range of vegetation and terrain conditions. Table 1 details the topographic and vegetation characteristics of each station. UAS-lidar was used to derive both a Digital Elevation Model (DEM) of the ground surface and a Digital Surface Model (DSM) of the ground surface with vegetation. This was done by first classifying ground returns using GreenValley International LiDAR360 software. The difference between these models (DSM – DEM) was then used to calculate the canopy height (CH). Canopy cover (CC) was calculated using the percentage of non-ground lidar returns relative to total returns. Leaf area index (LAI) was calculated using the approach of Richardson *et al.* (2009). Distance to the canopy edge (DCE) was calculated using the Mazzotti *et al.* (2019) approach with a vegetation cut height of 4 m. Northness (cosine of aspect), slope, and the topographic position index (TPI; Komarov and Sturm, 2023) were calculated using the MATLAB TopoToolbox software (Schwanghart and Scherler, 2014) and the UAS-lidar DEM. Shade factors were produced using the bare earth DEM and the canopy height model (CHM) separately to quantify the contribution of shading from vegetation and terrain. Using the TopoToolbox 'castshadow' function and solar position information for the study area from NREL's Solar Position Algorithm (SPA; Reda and Andreas, 2008), the total shaded hours were calculated considering the study period (January 5 – April 15, 2024). These values were normalized (0-1) based on the site-wide range. All automated stations were surveyed using a high-precision RTK GPS (<20 cm accuracy). Surveys were conducted using an Emliid RS2+ base station and Emliid RS3 rover. Surveyed coordinates were used to extract the nearest metric values (Table 1).

Table 1. Select UAS-lidar derived vegetation and terrain metrics near each station (Figure 1).

Metric	Site									
	CRN	1-OB	2-OG	3-OH	4-OM	5-CL	6-FN	7-FS	8-FE	9-SF
Canopy Height (CH, meters)	0.9	0.3	0.3	0.3	0.1	0.3	21.4	25.4	27.2	8.0
Canopy Cover (CC)	3%	0%	0%	0%	0%	0%	78%	91%	86%	12%
Leaf Area Index (LAI)	0.1	0.0	0.0	0.0	0.0	0.0	3.3	4.9	4.1	0.3
Distance (meters) to Canopy Edge (DCE)	80	101	95	65	16	4	-4	-2	-8	1
Northness	0.51	0.34	0.37	0.25	0.85	0.21	0.87	0.12	0.60	0.46
Slope	0.06	0.04	0.03	0.04	0.05	0.07	0.10	0.14	0.03	0.05
Topographic Position Index (TPI)	0.46	-0.04	-0.06	-0.01	-0.04	-0.04	0.17	-0.13	0.02	0.01
Topographic Shade Factor (DEM_{SF}, 0-1)	0.58	0.55	0.57	0.55	0.66	0.57	0.77	0.65	0.67	0.65
Canopy Shade Factor (CHM_{SF}, 0-1)	0.22	0.09	0.09	0.10	0.21	0.76	0.75	0.87	0.80	0.89

Stations were partitioned into unforested (CRN, #1-5) and forested areas (#6-9). Forested stations had varied forest densities from low (#9, CC = 12%) to high density (#7, CC = 91%). DCE had a wide range across all installed stations, from directly under the canopy (negative values) to more than 100 meters from the canopy edge (station 1-OB). North-facing slopes have a northness value of > 0.8 and south-facing of < 0.2. Slopes were gentle at all stations (3-14%), and the CRN station was noted as having the highest local prominence (TPI) of all stations. Terrain shading was noted to occur primarily at the 6-FN station, though there was a much larger range of shading due to vegetation at the stations (0.09 – 0.89) than due to terrain (0.55 – 0.77).

RESULTS AND DISCUSSION

Air and 5 cm soil temperatures at all installed stations were compared to observations from the CRN station (Table 2). For air temperature, distributions were unskewed and had similar standard deviations (5.9 – 6.4 °C). Differences in air temperature relative to the CRN site were minimal in all cases (< 1 °C) and temperature distributions had similar shapes. The soil temperature distributions had a range of standard deviations (2.0 – 3.6 °C) and were positively skewed to varied degrees (0.5 – 1.4), resulting in very different shapes. The standard deviations at forested and CRN sites were notably lower (2.0 – 2.5 °C) than open sites (3.1 – 3.6 °C). This suggests the range of soil temperatures was dampened due to increased vegetative cover. Sites with minimal vegetative cover (1-OB, 2-OG, 4-OM) and/or were less exposed (5-CL, 6-FN, 8-FE) tended to have colder soil temperatures (-0.2 to -1.1 °C) than were observed at the CRN station. As a result, the number of frozen soil days ranged widely (0 – 38 days). At several sites, frozen soil conditions were not observed throughout the study period (CRN, 7-FS, 8-FE, 9-SF). At the forested station with increased terrain shading (6-FN), soils were frozen for 20.1 days. At exposed open stations with less vegetative insulation (1-OB, 2-OG, 3-OH, 4-OM), frozen days ranged from 23.3 to 28.8 days. At the station with minimal vegetation and high shading (5-CL), frozen soils persisted for 38 days. The combination of high shading from the forest canopy and low interception (no canopy cover) resulted in ideal conditions for soil frost development and increased snow persistence at site 5-CL.

Table 2. Statistics of time-matched meteorologic observations at all Kingman Farm meteorological stations (Figure 1). Bias and the two-sample Kolmogorov-Smirnov test statistics (KS2 stat) are computed for each station relative to the CRN station. Frozen days are calculated as the total duration of time (in days) with 5 cm soil temperatures below 0 °C.

Site	air temperature (°C)					5 cm soil temperature (°C)					
	mean	std	skew	bias	KS2 stat	mean	std	skew	bias	KS2 stat	frozen days
CRN	1.1	5.9	0.0	--	--	3.1	2.5	0.5	--	--	0
1-OB	1.9	6.2	0.0	0.8	0.07	2.9	3.6	0.9	-0.2	0.38	28.8
2-OG	1.3	6.1	0.0	0.2	0.03	2.9	3.1	0.6	-0.3	0.37	28.4
3-OH	0.5	6.4	0.0	-0.6	0.08	3.1	3.5	0.8	0.0	0.31	24.7
4-OM	0.5	6.4	0.0	-0.6	0.08	2.6	3.1	0.9	-0.5	0.38	23.3
5-CL	0.6	6.2	0.0	-0.5	0.07	2.0	3.2	1.4	-1.1	0.51	38.0
6-FN	0.5	6.3	0.1	-0.7	0.09	2.2	2.4	1.0	-0.9	0.20	20.1
7-FS	0.4	6.0	0.1	-0.7	0.09	3.0	2.1	0.5	-0.1	0.13	0
8-FE	0.4	6.1	0.0	-0.7	0.09	2.8	2.0	0.5	-0.3	0.16	0
9-SF	0.5	6.3	0.1	-0.7	0.08	3.1	2.3	0.7	0.0	0.16	0

The time series of snow depth is shown for stations 1-9 in Figure 2. Seven measurable snow accumulation events were observed. For three snowfall events, snow persisted for less than ten days. The remaining four events occurred mid-winter when snow persisted from the middle of January to late February. Snow depth did not exceed 25 cm at any station. Snow cover duration at the stations ranged from 36.1 days (1-OB) to 63.5 days (5-CL). Descriptive statistics of snow cover and depth are included in Table 3. Station 5-CL was identified as having the most persistent snow, while the south-facing forested site (7-FS) had the shallowest snow. Increased snow cover duration (SCD) was linked to decreases in 5 cm soil temperatures ($R = -0.85$) at *in situ* stations. DCE was shown to be related to both snow persistence and depth observations (Table 4). Meanwhile, canopy density (CC, LAI) was linked to snow depth metrics (MSD, MMSD), and canopy shading (CHMSF) was linked to snow persistence (SCD, CSS). Table 4 presents the correlation results of station snow statistics to various metrics.

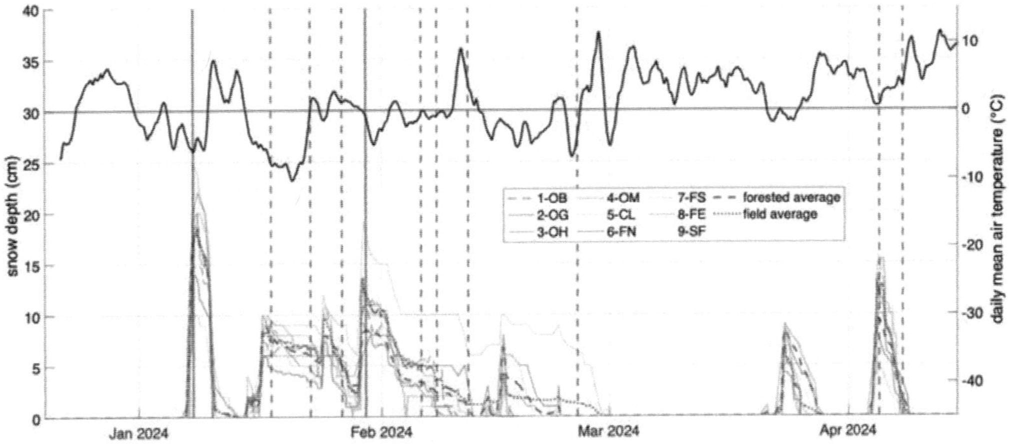

Figure 2. Snow depth (cm) observations from all *in situ* monitoring stations. Solid vertical lines indicate dates when UAS-lidar and optical sensors were flown, whereas dashed vertical lines indicate optical only. The CRN daily average air temperature is shown by the solid black line.

Table 3. Summary of snow condition statistics at all stations (except CRN) and correlations of the snow statistics to terrain and vegetation variables (Table 1).

Site	snow statistics			
	SCD (days)	CSS (days)	MSD (cm)	MMSD (cm)
1-OB	36.1	25.3	6.5	13.7
2-OG	36.5	26.9	4.4	9.9
3-OH	39.7	26.3	5.7	11.0
4-OM	43.2	32.6	6.1	11.9
5-CL	63.5	45.9	8.2	13.7
6-FN	56.6	39.4	5.0	9.3
7-FS	44.8	27.8	3.8	7.8
8-FE	48.3	28.0	5.5	10.9
9-SF	44.7	28.1	6.9	12.9

Table 3. Spearman's rank correlations, snow statistics (Table 3) to terrain and vegetation variables (Table 1).

Metric	Spearman's rank correlations			
	SCD	CSS	MSD	MMSD
Canopy Height (CH, meters)	0.2	0.1	-0.4	-0.6
Canopy Cover (CC)	0.3	0.2	-0.5	-0.6
Leaf Area Index (LAI)	0.3	0.2	-0.5	-0.6
Distance (meters) to Canopy Edge (DCE)	-0.4	-0.6	0.3	0.5
Northness	-0.4	0.4	0.0	-0.1
Slope	0.1	0.5	0.0	-0.2
Topographic Position Index (TPI)	-0.2	0.3	0.2	0.1
Topographic Shade Factor (DEM_{SF}, 0-1)	0.1	0.6	-0.3	-0.5
Canopy Shade Factor (CHM_{SF}, 0-1)	0.7	0.5	0.1	-0.1

Spatially distributed UAS snow depth and persistence observations were then used to assess the representativeness of the *in situ* stations (point sites) and to explore relationships between highlighted metrics and snow conditions site-wide. Interactions between forested and open areas resulted in high snow persistence in small forest clearings and the deepest snow in wind-sheltered areas near the south-facing forest-field boundary (Figure 3). In open areas, average UAS snow depths varied widely (~5 - 20 cm). Once the canopy cover fraction (CC) exceeded 0.6 (60%), a distinct decreasing trend in snow depth due to canopy interception was observed (Figure 4A). The highest snow depth variability was observed at the interface between forested and open areas (DCE < 25 m, Figure 4B). Depth then generally decreased until approximately 100 meters from the forest edge. At this point, the influence of the canopy on snow depth was negligible. Relationships between snow persistence, canopy shading, and DCE were highly variable (Figure 4C-D). The representation of interactions between various surface features is not trivial, as similar DCE can be associated with a range of snow conditions. For example, low DCE and high solar exposure resulted in the shallowest, least persistent snow. However, when paired with low solar exposure, low DCE can be used to identify the most persistent snow areas. Moderate snow persistence was observed in open areas (0.3 – 0.7), while areas experiencing high canopy shading had a wide range of observed persistence (0.1 – 1.0). The highest persistence (1.0) was prevalent in small, sheltered forest clearings, while the lowest persistence was observed primarily in areas with high canopy interception.

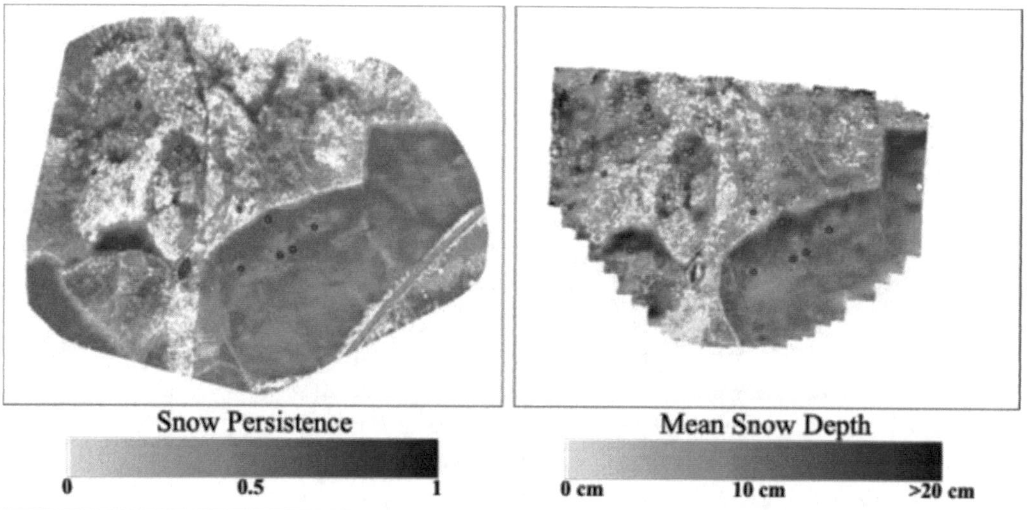

Figure 3. UAS observed snow persistence (left) and mean snow depth (right). The ground was not visible in UAS-optical imagery in areas of high vegetation density (white). Mean snow depth considers two flight dates (January 8 and January 30, 2024). Red markers indicate *in situ* stations (Figure 1).

While differences in mean air temperature at the local scale were generally small (< 1 °C), soil temperatures and soil freeze/thaw status varied widely across the small (~0.5 km^2) Kingman Farm study area. Based on limited UAS and continuous station observations, we estimate that seven snow microclimates exist at the site. While terrain influences cannot be ignored, canopy influences were identified as a primary control on the development of these microclimates at Kingman Farm. These included (1) open, high exposure (moderate depth, low persistence); (2) open, low-moderate exposure (moderate depth, moderate persistence); (3) forest gaps, low exposure (moderate-deep, high persistence); (4) forest-edge, low-moderate exposure (shallow-moderate depth, moderate persistence); (5) forest-edge, high exposure (shallow-moderate depth, low persistence); (6) dense forest, low exposure (shallow depth, moderate persistence); and (7) dense forest, moderate exposure (shallow depth, low persistence). Open areas are defined as far enough from the canopy that interactions with the forest are negligible (~100 m), forest gaps as areas with low canopy density surrounded by forests, and forest edges include the boundary between forests and open areas. Dense forests primarily include dense conifers, while deciduous forests at the site fall into the forest-edge classes. Exposure primarily refers to solar exposure, but wind exposure may be relevant depending on event-specific wind conditions. Wind speed and direction may influence the size of the forest edge area influenced by canopy effects. The *in situ* network of stations captured a range of forest densities (Figure 4A) and exposure regimes (Figure 4C).

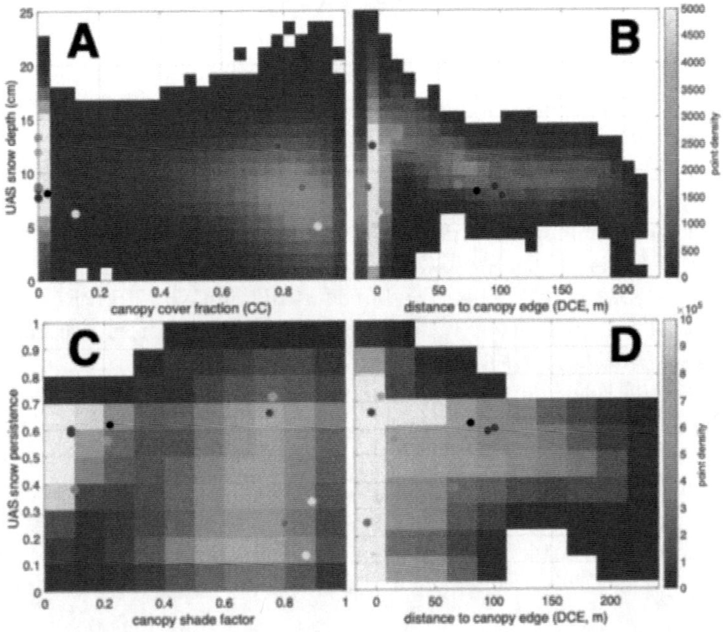

Figure 4. Density plots of UAS snow observations against derived site-wide metrics. Values at *in situ* stations (mean of 0.5 m buffer) are superimposed on each density plot, using the same marker colors as in Figure 1.

The UAS mean snow depth in the 1st and 3rd quartiles was 7.1 cm and 11.2 cm, while the station minimum and maximum were 4.9 cm and 13.3 cm. For UAS persistence (0 - 1), stations ranged from 0.13 to 0.72. The stations captured the range of normal conditions but did not capture snow depth or persistence outliers. Repositioning similar open sites (1-OB, 2-OG, 3-OH) to forest edges with high solar exposure, areas of intermediate canopy density, and locations further from the forest boundary may better capture site-wide spatial variability in snow conditions. Future work includes identifying a subset of land surface metrics that effectively represent snow spatial variability at sub-hillslope scales and creating a snow microclimate classification model using these critical predictors. This work highlights the limitations of using single meteorologic stations to represent surface conditions and informs optimal site placement for capturing hillslope-scale variability in snow conditions.

ACKNOWLEDGEMENTS

We thank students Imran Khan and Ryan Reed, who supported field sampling activities, and UNH land managers Stephen Eisenhaure, Evan Ford, and Peter Davis for enabling us to work at Kingman Farm. We also thank the Broad Agency Announcement Program and the Cold Regions Research and Engineering Laboratory (ERDC-CRREL) for funding and supporting this work. Distribution A: Approved for Public Release. Distribution is Unlimited. This material is based upon work supported by the Broad Agency Announcement Program and the Cold Regions Research and Engineering Laboratory (ERDC-CRREL) under Contract No. W913E523C0004

REFERENCES

Clark MP, Hendrikx J, Slater AG, Kavetski D, Anderson B, Cullen NJ, Kerr T, Örn Hreinsson E, Woods RA. 2011. Representing spatial variability of snow water equivalent in hydrologic and land-surface models: A review. *Water Resources Research*, **47**. *https://doi.org/10.1029/2011WR010745*

Johnston J, Jacobs JM, Cho E. 2023. Global snow seasonality regimes from satellite records of snow cover. *Journal of Hydrometeorology,* **25**: 65–88. *https://doi.org/10.1175/JHM-D-23-0047.1*

Judd Communications, n.d. Judd Communications Snow Depth Sensor. *http://static1.1.sqspcdn.com/static/f/1146254/15414722/1322862508567/ds2manual.pdf*

Kendall MG. 1948. *Rank Correlation Methods*. Griffin, London.

Komarov A, Sturm M. 2023. Local variability of a taiga snow cover due to vegetation and microtopography. *Arctic, Antarctic, and Alpine Research*, **55**: 2170086.

Massey Jr. FJ. 1951. The Kolmogorov-Smirnov test for goodness of fit. *Journal of the American Statistical Association*, **46**: 68–78. *https://doi.org/10.1080/01621459.1951.10500769*

Mazzotti G, Currier WR, Deems JS, Pflug JM, Lundquist JD, Jonas T. 2019. Revisiting snow cover variability and canopy structure within forest stands: insights from airborne lidar data. *Water Resources Research*, **55**: 6198–6216.

Reda I, Andreas A. 2008. *Solar Position Algorithm for Solar Radiation Applications (Revised)* (No. NREL/TP-560-34302, 15003974). *https://doi.org/10.2172/15003974*

Richardson JJ, Moskal LM, Kim S-H. 2009. Modeling approaches to estimate effective leaf area index from aerial discrete-return LIDAR. *Agricultural and Forest Meteorology*, **149**: 1152–1160.

Schwanghart W, Scherler D. 2014. Short Communication: TopoToolbox 2 – MATLAB-based software for topographic analysis and modeling in Earth surface sciences. *Earth Surface Dynamics*, **2**: 1–7.

Sturm M, Liston GE. 2021. Revisiting the global seasonal snow classification: an updated dataset for Earth system applications. *Journal of Hydrometeorology*, **22**: 2917–2938. *https://doi.org/10.1175/JHM-D-21-0070.1*

An Overview of NASA SnowEX Community-Collected Snow Pit Measurements (2017-2023)

MEGAN MASON[1,2], CARRIE VUYOVICH[1], KELLY ELDER[3], SVETLANA STUEFER[4], DRAGOS VAS[5], AND HANS-PETER MARSHALL[6]

ABSTRACT

Accurate assessment of snow characteristics is vital for understanding Earth's water, energy, and carbon cycles, as well as for effective water resource management and snow-related hazard mitigation. Despite advancements in satellite observations, critical gaps persist in global snow water equivalent (SWE) measurements. In this presentation, we present an overview of the primary ground-based snow measurements collected through the collaborative efforts of NASA SnowEx to provide extensive, requisite calibration and validation data sets to further mature various snow remote sensing techniques. We show snow depth, temperature, density, stratigraphy, grain size, wetness, liquid water content, and SWE at the point scale across a range of snow climate classes and vegetation types as measured from a snow pit. The snow pit measurements underscore the significance of well-refined manual ground-based techniques in complementing remote sensing technologies, particularly in areas where observations may be limited or insufficient. Through meticulous fieldwork by numerous participants, we have gathered comprehensive data sets across the suite of SnowEx field campaigns that offer insights into snow characteristics across different timescales, varied terrain, and regional processes that differ from montane forests to tundra landscapes. These measurements play a pivotal role in validating and refining remote sensing methodologies, ultimately contributing to the development of more accurate snow monitoring systems. Moreover, our presentation addresses the collaborative nature of the SnowEx initiative, which brings together national and international partners to collectively address challenges in snow monitoring. By synthesizing the collective volume of more than 1200 snow pits, we aim to contribute to the ongoing momentum to improve global snow observations.

---

[1] Hydrological Science Laboratory, NASA Goddard Space Flight Center, Greenbelt, MD, USA
[2] Science Systems Applications Inc. Lanham, MD, USA
[3] U.S. Forest Service, Fraser, CO, USA
[4] University of Alaska Fairbanks, Fairbanks, AK, USA
[5] Cold Regions Research and Engineering Laboratory, Fairbanks, AK, USA
[6] Boise State University, Boise, ID, USA
Corresponding author: megan.a.mason@nasa.gov

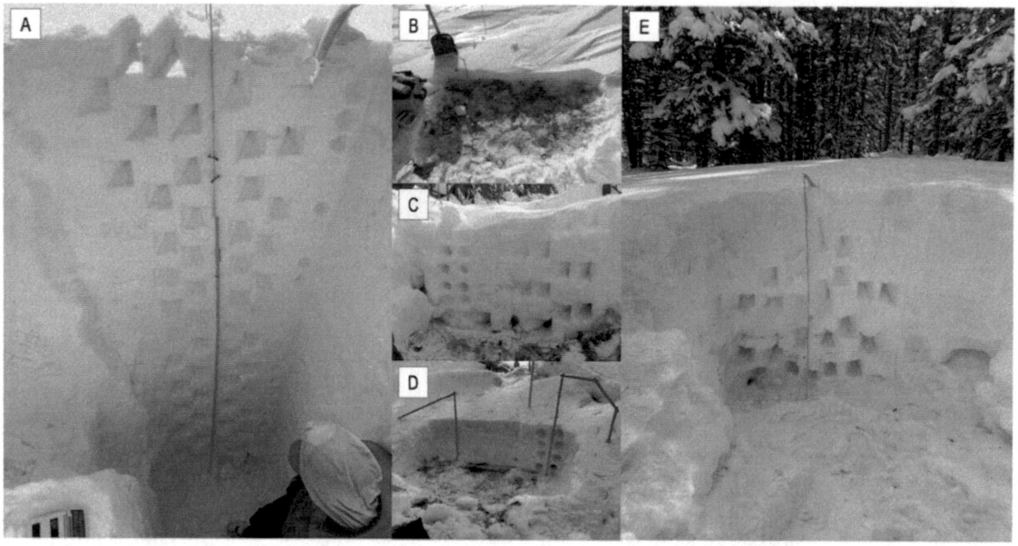

Figure 1. Snow pit images taken as part of the standard sampling protocol throughout the many NASA SnowEx field campaign sampling periods. A) Example of a deep (>2 m), late season snow pit in Little Cottonwood Canyon, UT. B, C) Both snow pits are part of the SnowEx 2023 Tundra (B) and Boreal Forest (C) Campaign in Alaska. They show the distinct surface wind effects and undulating ground surface, respectively. D, E) Snow pits that represent two (of six) different snow classes: Maritime at Sagehen Creek, CA (D) and Montane Forest at Niwot Ridge, CO (E), with both study areas under a forested canopy.

Northeast Snow Survey (NESS): A Feasibility Study to Develop an Automated Snowpack and Climate Monitoring Network in the Northeast United States

ELIZABETH BURAKOWSKI[1], JOSHUA BENEŠ[2], JORDAN CLAYTON[3], ALIX CONTOSTA[1], ANNA GRUNES[2], HEATHER HOFMAN[4], CARA MCCARTHY[4], SCOTT MCKIM[5], GEORGIA MURRAY[6], CHRIS NADEAU[7], SARAH NELSON[6], KYLER PHILLIPS[6], ERIKA ROWLAND[8], AND MELISSA WEBB[4]

ABSTRACT

Winter is the fastest warming season East of the Mississippi River, with winter warming hotspots (greater than +2.5 ºC since 1970) concentrated in the northeastern United States. Over the past 100 years, the region has lost 2-3 weeks of sustained winter conditions across low elevations and some high montane areas. Most of what we know about winter climate change in the northeastern US comes from low-elevation weather stations in populated areas, with a scarcity of snowpack and weather observations across elevational gradients in more remote locations. This differs from the western U.S., where the Natural Resources and Conservation Service Snow Telemetry (SNOTEL) network monitors snowpack, weather, and other climate elements at over 900 automated stations and provides critical data for hydrological monitoring and water resource management. The Northeast Snow Survey (NESS) Feasibility Study was funded in the FY 2023 federal omnibus bill to identify interest-holder needs for snowpack and weather monitoring and design a station network with a supporting maintenance and operations plan for the northeastern U.S. to meet those needs. During the first phase of the feasibility study, we have solicited input from interest-holders via surveys, workshops, and interviews. Respondents represented diverse interests including researchers (climate, ecology, wildlife), forecasters (hydrology/flood, weather, avalanche), land managers (forestry, agriculture, wildlife, infrastructure, outdoor recreation), environmental educators and individual recreationists. We are currently conducting a data survey to collate all relevant sources of snowpack and weather data in the region to assess spatial gaps, identify high priority areas for potential network station locations, and evaluate the suitability of candidate locations for fulfilling network objectives. We look forward to hearing from Eastern Snow Conference attendees about how this northeastern automated snowpack and climate monitoring network could support their research, forecasting, management, and other interests.

[1] University of New Hampshire, Durham, NH USA
[2] University of Vermont, Burlington, VT USA
[3] USDA NRCS Utah Snow Survey, Salt Lake City, UT USA
[4] USDA NRCS National Water and Climate Center, Portland, OR USA
[5] New York State Mesonet and SUNY-Albany, Wilmington, NY USA
[6] Appalachian Mountain Club, Gorham, NH USA
[7] Schoodic Institute, Winter Harbor, ME USA
[8] Forest Stewards Guild, Bridgton, ME USA
Corresponding author: jbenes@uvm.edu

Examination of Microwave Backscatter of Freshwater Lake Ice using Polarimetric Decomposition

CONNOR MCRAE-PHARO[1] AND GRANT GUNN[1]

ABSTRACT

Polarimetric decomposition emerges as a critical technique for the analysis of microwave signals reflected by freshwater ice features. This study leverages RADARSAT-2 quad-polarization captures to examine the backscatter characteristics from Old Crow Lake in Yukon and Noel Lake in the Northwest Territories, illustrating distinct morphometric contrasts (shallow and deep, respectively). This differentiation in depth facilitates the exploration of unique ice compositions, particularly focusing on the presence or absence of tubular bubbles at the ice-water interface. Through polarimetric decomposition, we identify the predominant scattering mechanisms in ice with and without tubular bubbles. Our findings reveal that the signal from both locations predominantly exhibits single-bounce scattering, independent of lake morphometry, with significantly weaker volume-scattering signals and minimal to absent double-bounce signals. These observations are of substantial significance for remote sensing applications, including the potential for enhanced freshwater ice thickness retrieval, as interferometry achieves greater accuracy with single-bounce interactions. Our research underscores the utility of polarimetric decomposition in analyzing freshwater ice backscatter, demonstrating its potential to enrich our understanding of ice's physical properties and their impact on radiative transfer. This method holds promise for advancing current ice and climate monitoring techniques in the cryosphere.

[1] University of Waterloo, Waterloo, ON, Canada
Corresponding author: cmcraeph@uwaterloo.ca

Observations of Precipitation Phase in the Tropical High Andes of Bolivia and Peru

BELLA VANDEN BOOM[1], L. BAKER PERRY[1], MAXWELL RADO[2], MARCOS ANDRADE[3], LAURA TICONA[3], AND SANDRO ARIAS[4]

ABSTRACT

High-elevation glacierized cordilleras of the tropical Andes of southern Peru and Bolivia serve as significant water towers that sustain communities and ecosystems downstream. Recent climate change associated with increasing precipitation variability, rising freezing levels, and a higher percentage of precipitation falling as rain has accelerated glacier retreat across the region, exacerbating concerns about future freshwater availability for natural systems, agriculture, hydroelectric generation, and other human activities. This poster investigates methods to partition precipitation phase (e.g., snow, graupel, mixed rain/snow, rain) – a critical control on glacier behavior – using in situ observations from Laguna Sibinacocha (4,895 m) in the Cordillera Vilcanota of Peru and Nevado Chacaltaya (5,160 m) in the Cordillera Real of Bolivia. Precipitation phase was recorded at the top of each hour by an OTT Parsivel present weather sensor and an OTT Pluvio2 weighing precipitation gauge recorded hourly liquid equivalent precipitation totals. The frequency and total liquid equivalent partitioned by precipitation phase were analyzed according to wet and dry bulb air temperature. Initial results indicate that 8% of all precipitation at Laguna Sibinacocha was identified by the OTT Parsivel as rain, with approximately 89% of precipitation identified as snow and graupel. Succeeding results at Nevado Chacaltaya indicate that 3% of all precipitation was categorized by the OTT Parsivel as rain, with approximately 95% of precipitation categorized as snow and graupel. These findings will improve partitioning of precipitation phase by temperature-based methods at other tropical high mountain regions where observations of precipitation phase are not available.

[1] Department of Geography and Planning, Appalachian State University, Boone, NC, USA
[2] Universidad Nacional de San Antonio Abád del Cusco, Cusco, Peru
[3] Atmospheric Physics Laboratory, Universidad Mayor de San Andrés, La Paz, Bolivia
[4] Servicio Nacional de Meteorología e Hidrología, Cusco, Peru
Corresponding author: vandenboomir@appstate.edu

Photogrammetric Historical DEM Generation to Support the Mapping of Glacier Area, Length, and Volume Change in the Canadian High Arctic

CAROLYN GORWILL[1] AND LAURA THOMSON[1]

ABSTRACT

The glaciers of the Canadian Arctic are currently the third largest contributor to global sea level rise due to the amplification of Arctic warming which drives increased glacier melt and runoff. Long-term glacier records in the Canadian High Arctic are extremely valuable in understanding past glacier changes and informing projections of future change in the face of anthropogenic climate change. A detailed inventory of glacier coverage within Axel Heiberg Island's glacial hydrological basins based on historical air photos from 1958-59 exists and was compared to new glacier extents mapped from 1999–2000 satellite imagery, revealing that glacier retreat accounted for ~60 km^2 of area loss in the 42-year period. This retreat is likely in response to warming temperatures, with Eureka, NU, east of Axel Heiberg Island, NU, experiencing a 3.2 °C increase in annual average surface temperature since 1972.

While the past mapping of historic glacier coverage on Axel Heiberg Island provides valuable insight into regional glacier change, historic information on individual glacier areas, lengths and surface elevations in the region is lacking. With photogrammetry, the National Air Photo Library's collection of historical photographs over Axel Heiberg Island can be used to construct historical digital elevation models of glaciers to ascertain ice thickness changes in the region. This study therefore presents the workflow and case studies of historic DEM and orthomosaic creation using historical imagery from the late 1950s, and high-resolution contemporary satellite imagery (3m) and digital elevation models (2m) which provide ground reference for model building. These new historic DEMs, in combination with individual glacier extents derived from the orthorectified imagery will allow for a more comprehensive assessment of glacier change since the late 1950s.

Figure 1. Historical DEM production workflow over Western Ice Cap, Axel Heiberg Island. (a) alignment of NAPL historical imagery (b) Digital elevation model creation (9.75m) (c) Orthorectified imagery.

[1] Department of Geography and Planning, Queen's University, Kingston, ON, Canada
Corresponding author: 19crg4@queensu.ca

Assessing Uncertainty in Cosmic Ray Neutron Sensor Snow Water Equivalent Measurements from Heterogeneous Snow Distributions

HAEJO KIM[1], SAM TUTTLE[2], AND ERIC SPROLES[3]

ABSTRACT

Accurate monitoring of Snow Water equivalent (SWE), the amount of water stored in snowpack, for shallow snowpack is complicated in many regions. For cold windy environments like the Montana prairies, snow is redistributed from open areas and accumulates around barriers (Such as wind breaks or crop stubble) and low-lying areas, leading to extreme variation in snow depths. conventional methods of monitoring SWE, such as snow pillows and snow depth sensors, are not reliable for accurately representing prairie SWE. Cosmic ray neutron sensing (CRNS) has shown promise in being able to estimate SWE continuously in such heterogeneous conditions by integrating across a large areal footprint (~200 m radius). In winter 2020-2021 we deployed a CRNS sensor at the Central Agricultural Research Center (CARC) in Moccasin, Montana as part of the NASA SnowEx Prairie field campaign, and collected UAV lidar and ground observations. Early results show that including in-situ snowpack conditions from a UAV lidar survey in a neutron transport model can account for the effects of bare ground and heterogeneous snow distributions on CRNS SWE observations. This analysis aims to improve our understanding of the major factors that influence CRNS measurements. How does the surrounding snow heterogeneity around a CNRS sensor bias SWE estimates and what recommendations can be made in siting future CNRS sensors for snow monitoring in similar locations? Through this work, we hope that CRNS will be a tool to benchmark future satellite missions for National and global snow monitoring.

[1] Syracuse University, Syracuse, NY, USA
[2] Syracuse University, Syracuse, NY, USA
[3] Montana State University, Bozeman, MT, USA
Corresponding author: hkim139@syr.edu

Figure 1. Maintenance done on cosmic ray neutron sensor by Dr. Eric Sproles and Dr. Sam Tuttle at the Central Agricultural Research Center in March 2024. Photo taken by Haejo Kim.

Oral Session #2

Session Chairs: Carrie Vuyovich & Aleksandra Elias Chereque

Reducing a complex multi-layered snowpack to a radar equivalent snowpack with minimal number of layers

JULIEN MELOCHE[1], NICOLAS LEROUX[2], BENOIT MONTPETIT[1], VINCENT VIONNET[2], AND CHRIS DERKSEN[1]

ABSTRACT

New efforts for spaceborne radar-based SWE retrieval algorithms are under development and the methodology and science behind these retrievals is improving. A state-of-the-art SWE retrieval needs a radiative transfer model (RTM) to link snow properties to backscatter. Snow properties like layer thickness, density, temperature, and microstructure are necessary to properly model the microwave signal with an RTM. To improve the computational cost of running an RTM, there is a need to convert a multi-layered snowpack with large number of layers (20-50) into a simplified snowpack (2-3 layers) that preserves its snow mass and scattering behavior. The goal of this work was to develop a simple algorithm to obtain a radar equivalent snowpack that preserves the snow mass and scattering behavior. Layers were grouped and averaged based on the thickness and the microwave extinction coefficient of the layers. To evaluate our method, we tested it on multi-layered CROCUS simulations at 11 sites across various snow climates over multiple seasons. From the n-layer simulations (maximum layering), we compare various snowpack reduction methods by evaluating differences in snow mass and simulated backscatter using the Snow Microwave Radiative Transfer Model.

[1] Climate Research Division, Environment and Climate Change Canada, ON, Canada
[2] Meteorological Research Division, Environment and Climate Change Canada, QC, Canada
Corresponding author: julien.meloche@ec.gc.ca

Towards Understanding the Impact of Spatial Resolution and Viewing Geometry on Passive Microwave Snow Retrievals

LINA ZSCHENDERLEIN[1,2], RICHARD KELLY[1], KARI LUOJUS[2], CHRIS DERKSEN[3], AND COLLEEN MORTIMER[3]

ABSTRACT

Passive microwave data of high spatial resolution—stemming from enhanced-resolution data products and future satellite missions—are considered beneficial for the performance of geophysical snow parameter retrievals. However, the precise implications of spatial resolution and related viewing geometry effects on snow retrieval methodologies remain unclear. We therefore investigate whether high spatial resolution passive microwave data can indeed capture the very high spatial natural variability of snow conditions, and whether the sensor viewing geometry can further help to resolve the spatial variability of land surface features. The objectives are to quantify the impact of spatial resolution on snow water equivalent (SWE) retrieval accuracy and uncertainty, and to examine the relevance of look angle variations on snow detectability. To estimate SWE across Canada and the US, selected MEaSUREs Calibrated Enhanced-Resolution brightness temperature data (19 and 37 GHz) of varying spatial resolution (25 km, collocated 6.25 and 3.125 km) serve as input for the GlobSnow SWE retrieval framework. To map snow cover extent, fore and aft-looking WindSat brightness temperature data (19 and 37 GHz) are used for the EUMETSAT H SAF snow detection algorithm, offering an almost instantaneous view of the same ground scene from two different look angles. Independent reference data include gamma SWE observations and snow cover maps by the Interactive Multisensor Snow and Ice Mapping System (IMS), amongst others. By addressing the role of topography, landcover types, and *in situ* snow measurement distribution, we aim to provide a comprehensive understanding of the scaling and viewing nature of satellite data for snow retrieval and mapping purposes. This study is expected to demonstrate the potential of dual-view passive microwave data of high spatial resolution to advance terrestrial snow monitoring, in preparation of the future Copernicus Imaging Microwave Radiometer (CIMR) mission.

[1] University of Waterloo, Waterloo, ON, Canada

[2] Finnish Meteorological Institute, Helsinki, Finland

[3] Environment and Climate Change Canada, Toronto, ON, Canada

Corresponding author: lina.zschenderlein@uwaterloo.ca

Sensitivity Study of 13.5 GHz Ku Backscatter and Snow Depth at a Farm Site in Powassan, Ontario

WEI WANG[1], RICHARD KELLY[1], AND PETER TOOSE[2]

ABSTRACT

Proposals like CoReH2O and TSMM together with various airborne and ground campaigns conducted in support have shown the potential of SWE retrieval using active microwave and cost functions based on snow radiative transfer models. However, there are still uncertainties in the choices of frequency combination in addition to needs to conduct more experiment covering different classes of snowpacks using airborne systems. As part of the efforts to fill the current gaps, the CryoSAR winter 2022-23 campaign was conducted using an advanced Ku (13.5 GHz) and L (1.3 GHz) band airborne SAR system supported by ground *in situ* measurement and Lidar drone data at a farm site in Powassan, Ontario. The airborne system flew eight flight paths around the study area, providing good coverage and data with various incidence angles. This paper describes a preliminary sensitivity study aimed at demonstrating the correlation between snow depth and Ku band backscatter. The study area consists of open farm fields, wooded areas and some farm buildings. As forest will reduce the sensitivity of Ku backscatter to snow depth, this study will focus on open farm fields by first masking out the wooded areas and buildings. Snow depth over the study area is acquired by differencing bare ground and snow-on DEM, both of which were generated using Lidar drone data. Due to the relatively shallow snow depth, background surface scattering from snow-soil interface needs to be removed from the total backscatter, and this is accomplished through the addition of L band signal. A positive correlation between the Ku backscatter and snow depth is anticipated, and this study will lay a solid foundation for future SMRT modelling and SWE retrieval efforts.

---

[1] Department of Geography and Environmental Management, University of Waterloo, Waterloo, ON, Canada.

[2] Climate Processes Branch, Environment and Climate Change Canada, Toronto, ON, Canada

Corresponding author: w444wang@uwaterloo.ca

Multi-Physics Ensemble Modelling of Arctic Tundra and Taiga Snowpack Properties

GEORGINA J. WOOLLEY[1], NICK RUTTER[1], LEANNE WAKE[1], VINCENT VIONNET[2], CHRIS DERKSEN[3], RICHARD ESSERY[4], PHILIP MARSH[5], ROSY TUTTON[5], BRANDEN WALKER[5], MATTHIEU LAFAYSSE[6], AND DAVID PRITCHARD[7]

ABSTRACT

Sophisticated snowpack models fail to simulate density and SSA profiles of Arctic snowpacks due to an underestimation of wind-induced compaction, misrepresentation of basal vegetation impacting compaction and metamorphism, and omission of water vapour transport. Parameterisations of Arctic snow physical processes were implemented into a 120-member ensemble version of the Soil, Vegetation and Snow version 2 (SVS2-Crocus) land surface model, creating Arctic SVS2-Crocus. Arctic SVS2-Crocus was driven and evaluated using measurements of snowpack properties (snow water equivalent, depth, density and SSA) at Trail Valley Creek (TVC), NWT, Canada over 32-years (1991-2023). The inclusion of wind-induced compaction in Arctic SVS2-Crocus increased surface snow density and reduced the RMSE by 41% (176 kg m^{-3} to 103 kg m^{-3}). Parameterisations of basal vegetation were less effective in reducing compaction of basal snow layers (default RMSE: 67 kg m^{-3}; Arctic RMSE: 65 kg m^{-3}). All 120 ensemble members of Arctic SVS2-Crocus outperformed default SVS2-Crocus simulations of snow density profiles at TVC, as evidenced by lower continuous ranked probabilistic scores.

The recent addition of a forest canopy model to Arctic SVS2-Crocus will be evaluated across a 40-km transect of the Northwest Territories, Canada, where tundra to taiga ecosystems transition. Quantifying the impact of forest and shrub canopies on simulated snow properties will be crucial for the evaluation of satellite microwave radiative transfer models for snow water equivalent retrieval (Terrestrial Snow Mass Mission, TSMM) and coupling of Arctic SVS2-Crocus with CryoGrid to evaluate simulated soil temperatures and their impact on permafrost.

[1] Department of Geography and Environmental Sciences, Northumbria University, Newcastle Upon Tyne, UK.

[2] Meteorological Research Division, Environment and Climate Change Canada, Dorval, QC, Canada.

[3] Climate Research Division, Environment and Climate Change Canada, Toronto, ON, Canada.

[4] School of Geosciences, University of Edinburgh, Edinburgh, UK.

[5] Cold Regions Research Centre, Wilfrid Laurier University, Waterloo, ON, Canada.

[6] Météo-France, CNRS, CNRM, Centre d'Études de la Neige, Universite Grenoble Alpes, Universite de Touloise, France.

[7] School of Engineering, Newcastle University, Newcastle Upon Tyne, UK

Corresponding author: georgina.j.woolley@northumbria.ac.uk

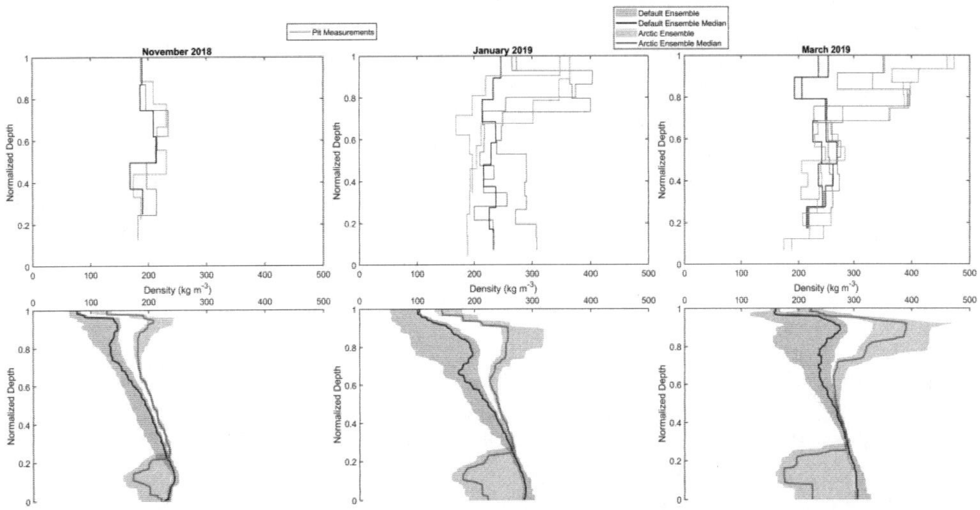

Figure 1. Comparison of measured and simulated vertical profiles of density (kg m^{-3}, median, interquartile range) by default and Arctic SVS2-Crocus from the November 2018, January 2019, and March 2019 winter field campaigns.

Oral Session #3

Session Chairs: Joan Ramage & Gifty Attiah

Bias-Correction of ERA5-Land and FLDAS Reanalysis-Based Soil Temperature using Machine Learning

TYLER C. HERRINGTON[1], ANDRE R. ERLER[2], AND CHRISTOPHER G. FLETCHER[1]

ABSTRACT

Reanalysis products provide spatially homogeneous coverage for a variety of climate variables, which is invaluable for regions where observational data are limited. ERA5-Land is a state of the art reanalysis system that provides high-resolution (9 km) estimates of critical land surface variables such as vertically-resolved soil temperature that is used to estimate active layer depth and permafrost extent. Recent research has shown that ERA5-Land soil temperatures show substantial warm biases over permafrost regions. Here we use mean bias subtraction (MBS), multiple linear regression (MLR) and random forest regression (RF) to perform bias correction of the ERA5-Land soil temperature product. The MLR and RF models employ 10 predictors, including soil depth, product soil temperature, information regarding the seasonal cycle of soil temperature, air temperature, vegetation, snow cover, elevation, latitude and longitude. The RF model substantially outperforms MBS and MLR over all regions and latitudes, providing an average RMSE reduction (relative to ERA5-Land soil temperature) of between 50% and 57% when the ground is snow-covered, and between 43% and 64% during the snow-free season. Most notably, the RF model reduces the RMSE over high-latitude regions north of 60N by 52% to 57% in the snow-covered season, when ERA5-Land soil temperatures exhibit an RMSE of > 4K. The bias-corrected soil temperature product provides gridded soil temperature data over the extratropical northern-hemisphere at a resolution of 5 km and will be useful for a wide range of applications, including as an initialization condition for hydrological models, and as a tool to validate model soil temperatures.

---

[1] Geography and Environmental Management, University of Waterloo, Waterloo, ON, Canada
[2] Aquanty Inc., Waterloo, ON, Canada
Corresponding author: therring@uwaterloo.ca

Determining Summer Accumulation and Seasonal Precipitation Sources and Pathways in the Western Queen Elizabeth Islands: Case study at White Glacier on Umingmat Nunnat (Axel Heiberg Island), NU

SOFIA GUEST[1], LAURA THOMSON[1], AND ALISON CRISCITIELLO[2]

ABSTRACT

Rapid changes occurring across Arctic glacial landscapes provide a clear indication of global climate change. The majority of the Arctic is experiencing a trend of increased temperature and precipitation, however, the 63-year mass balance record on White Glacier, Umingmat Nunnat (Axel Heiberg Island), NU, does not show any statistically significant trend in precipitation. Summer accumulation is currently not accounted for in mass balance measurements. In this study, the magnitude of summer snowfall and the isotopic composition of the snowpack of White Glacier and its surrounding area are analyzed to determine spatial and temporal variations and precipitation event sources and pathways. A Picarro cavity ring-down spectrometer is utilized to determine the ratios of stable water isotopes found in this snowpack. Precipitation event sources and pathways are ascertained through HYSPLIT modelling and source temperatures are found using the Simple Water Isotope Model. Preliminary findings indicate that summer accumulation could account for one-half to two-thirds of accumulation, significantly impacting mass balance. Additionally, the increase $\delta^{18}O$ found in snow and firm samples indicates a continued shift in driving modes of climate, potentially from enhanced meridional vapour flux.

[1] Queen's University, Kingston, ON, Canada
[2] University of Alberta, Edmonton, AB, Canada
Corresponding author: 22smg@queensu.ca

Assessment of ERA5-Land Lake Ice Related Variables from Satellite Observations

ARIANA MANSINGH[1], CLAUDE DUGUAY[1], AND GRANT GUNN[1]

ABSTRACT

Lake ice is a key component of the northern hydrologic and atmospheric systems, essential for characterizing climate. As the presence/absence of lake ice alters the heat and energy transfer of the lake-atmosphere interface, it has an impact on local and regional weather/climate variables such as air temperature and precipitation. Thus, accurate observations or modeling of lake ice conditions can improve estimates of related atmospheric variables. This research focuses on the assessment of lake ice-related variables provided in the European Centre for Medium-Range Weather Forecasts (ECMWF) ERA5-Land product: lake surface temperature, lake ice depth and albedo; which can inform potential improvements to the reanalysis product prior to its future release (i.e., ERA6). The assessment is conducted through a comparison with satellite-derived products, notably the Interactive Multisensor Snow and Ice Mapping System (IMS) Daily Northern Hemisphere Snow and Ice Analysis, and Moderate Resolution Imaging Spectroradiometer (MODIS) land surface temperature (LST) and albedo products. The IMS product provides the lake surface cover (ice and water) from which ice phenology dates (freeze-up start, freeze-up end, break-up start, break-up end) can be determined. ERA5-Land's lake ice depth estimates are used to derive comparable ice phenology dates for comparison with the IMS product. Assessment for both LST and albedo is performed similarly to each other, where ERA5-Land's estimates of lake ice surface temperature and forecasted albedo are compared to the MODIS LST and albedo products, respectively. Prior to comparison, both MODIS products are spatially aggregated from their 500 m grid spacing onto that of ERA5-Land at 9 km. This presentation will report on the magnitude and potential sources of biases in ERA5-Land to inform potential improvements to the FLake model embedded for the generation of lake-related variables in the reanalysis product.

[1] Department of Geography and Environmental Management, University of Waterloo, Waterloo, ON, Canada

Corresponding author: a2mansin@uwaterloo.ca

A Field Study: *In situ* Observations of Lake Ice Break-Up Processes

ARASH RAFAT[1], HOMA KHEYROLLAH POUR[1], CHRISTOPHER SPENCE[2], AND MICHAEL J. PALMER[3]

ABSTRACT

Measurements of freshwater ice thermodynamics during freeze-up and break-up have been remarkably limited, due in part to the inadequate load bearing capacity of thin or candled ice. Hence, many thermal ice models suffer from inaccurate or simplified thermodynamic parameterizations based on data collected from sea ice investigations. In this study, we present a complete record of *in situ* measurements of lake ice thermodynamics during break-up using a constructed Floating Research Station (FRS). The FRS is anchored in a 1 km^2 shallow subarctic lake, Landing Lake, located ~12 km north of Yellowknife, Northwest Territories, Canada in the Baker Creek Research Watershed. Ice thermodynamics are recorded using a Snow and Ice Mass Balance Apparatus (SIMBA). A meteorological station on Landing Lake records ambient environmental conditions. The significance of surface and bottom melt processes, light penetration through ice, and the role of mechanical decay is highlighted. Challenges on the use of floating rafts for monitoring break-ups processes in lakes are discussed, and areas of future improvement for subsequent studies identified. Outcomes aim to improve thermodynamic parameterizations in existing freshwater thermal ice models, while identifying difficulties associated with raft design.

[1] Cold Regions Research Centre and Department of Geography and Environmental Studies, Wilfrid Laurier University, Waterloo, ON, Canada

[2] Environment and Climate Change Canada, Saskatoon, SK, Canada

[3] North Slave Research Centre, Aurora Research Institute, Aurora College, Yellowknife, NT, Canada

Poster Session #2

Session Chair: Krystopher Chutko

Advancing Snow Modelling across Canada from the Arctic to Southern Regions

AGNES RICHARDS[1], FELIX OUELLET[1], ERIKA BOISVERT-VIGNEAULT[2,3], AND ALEXANDRE LANGLOIS[2,3]

ABSTRACT

The climate is changing rapidly across Canada and in some southern regions rain-on-snow events are becoming more common with slush dominating the roadways. We adapted the SNOWPACK model from traditional avalanche applications to the Canadian Artic, as well as for two southern regions with moderate (Bay of Quinte, Ontario) to low snowfall (Wigle Creek, Ontario). We developed a downscaling tool (Outil de Spatialisation de SNOWPACK pour l'Arctique - OSSA) using changes in slope, which refined the spatial resolution from 45 km² to 1 km². Our simulations in the Arctic demonstrated that icing events tripled across the Canadian Arctic Archipelago between 1979-2011. SNOWPACK simulations (1970s to 2020) for the Bay of Quinte focused on changes in snow parameters such as Snow Water Equivalent (SWE), which drives snow melt and flooding. For the Bay of Quinte and Wigle Creek we used North American Regional Reanalysis (NARR) data to drive SNOWPACK which a suite of weather parameters such as precipitation, radiations, humidity, wind, and air temperature. The major difference between the two southern regions was the calibration. For both regions air temperatures were tested at various levels of pressure. At the Bay of Quinte, a full calibration was obtained with air temperatures at 500 hPa. In contrast, at Wigle Creek the best calibration was obtained with surface air temperatures. To better understand these differences in calibrations a sensitivity analysis was performed on the precipitation phase. Further SNOWPACK validations were performed using field snow pit data for snow water equivalent, soil temperatures, and humidity.

[1] Government of Canada, Environment and Climate Change Canada, Burlington, Ontario, Canada

[2] Centre d'Applications et de Recherches en Télédétection (CARTEL), Université de Sherbrooke, Sherbrooke, QC, Canada

[3] Groupe de Recherche Interdisciplinaire sur les Milieux Polaires (GRIMP)

Corresponding author : felix.ouellet.ec.gc.ca

Remote Sensing and Cloud Computing: Creating Canada-Wide Lake Ice Phenology Products using Google Earth Engine and Sentinel-1 SAR Imagery

BRENDAN WARK[1]

ABSTRACT

Documenting lake ice phenology allows for a deeper understanding of the effects of climate change and more accurate climate modelling. Tracking lake ice phenology using *in situ* measurements is too costly over a large study area. This project uses remote sensing to quickly create large-scale lake ice phenology products at no cost to the user. All image processing is completed using cloud computing to further improve ease-of-use and lower hardware requirements.

The project creates melt and freeze onset dates using a dynamic threshold variability method (DTVM) algorithm on Google Earth Engine. The algorithm uses a shapefile of lakes as input, then collects all Sentinel-1 C-band synthetic aperture radar (SAR) acquisitions available over those lakes for a given year. The DTVM creates thresholds based on the backscatter of the SAR imagery and uses the thresholds to determine melt and freeze onset dates for each lake. This algorithm is run on 40 study areas across Canada, including over 2,000 lakes, and the results are combined in ArcGIS. An interpolation tool is used to generate melt onset, freeze onset, and lake ice season length rasters for all of Canada.

Algorithm validation has been conducted on 3 study areas with 500 lakes each. Each study area has a unique latitude. Using results from the Canadian Lake Ice Model (CLIMo), temperature records from weather stations, and DAYMET V4 daily temperature estimates, the algorithm can reliably determine melt and freeze onset dates across Canada.

[1] Department of Geography and Environmental Management, University of Waterloo, Waterloo, ON, Canada

Corresponding author: bwwark@uwaterloo.ca

Interferometric and Polarimetric Analysis of SAR Freshwater Lake Ice Acquisitions

JAKE E. FERGUSON[1] AND GRANT E. GUNN[1]

ABSTRACT

The physical structure of lake ice is an important feature of northern systems. Lake ice properties serve as important inputs into climatological models, are relevant drivers of hydrological and ecological phenomena, and impact human operations and safety in Canada's numerous lake-dense northern regions. While critical systems can be monitored in-situ, remote sensing derived data products are the only option for the ongoing monitoring of hundreds of thousands of isolated lakes.

Although Synthetic Aperture Radar (SAR) is attractive for freshwater ice observation due to its ability to selectively penetrate cloud, snow, and ice cover by modulating the wavelength employed, extracting ice thickness data remains difficult as observable properties of the returned wave are strongly ambiguous confluences of different system properties. Leveraging modern C-band SAR infrastructure and knowledge of natural ice scattering mechanisms, we investigate the potential of a differential interferometric SAR (DInSAR) approach to ice growth rate extraction as well as a polarimetric approach to ice structure extraction.

RADARSAT Constellation Mission (RCM) 5m compact-polarization acquisitions of Lhù'ààn Män' (Kluane Lake) in Yukon, Canada over the 2022-2024 winter seasons (January-March) are paired to generate interferograms. Topographic phase contributions are simulated and removed using ArcticDEM as the assumed topology, with the remaining phase interpreted as ice growth after accounting for refraction and wavelength compression. Average DInSAR derived ice growth rate shows reasonable agreement with thermodynamically-simulated values for the 2022-2024 seasons as well as in-situ values measured in the 2024 season. We also evaluate the potential of observing ice features including ridging and to-bed freezing using M-Chi compact polarization decomposition parameters.

[1] Department of Geography and Environmental Management, University of Waterloo, Waterloo, ON, Canada
Corresponding author: j5fergus@uwaterloo.ca

Comparing Four Monitoring Methods of Snow Depth at Dorset, Ontario, Canada

HUAXIA YAO[1], CHRISTOPHER MCCONNELL[1], RICHARD FERNANDES[2], ANDY BEATON[3], TIM FIELD[4], APRIL JAMES[4], AND STEVEN FASSNACHT[5]

ABSTRACT

Snow depth serves as a crucial indicator for understanding snow hydrology, water resources, and climate change. Various monitoring methods have been employed to assess snow depth, necessitating comprehensive comparisons to determine their efficacy. This study evaluates four distinct methods utilized at the Dorset climate station within the long-term inland waters monitoring network of Ontario, Canada, since 2010. The methods compared include Sonic Ranging Sensor (SRS) providing 10-minute data, Copernicus (GMES: Global Monitoring for Environment and Security) remote sensing, Time-Lapse Photography (GEOMATICS) with 30-minute intervals, and manual measurements conducted daily to weekly. Comparative analyses against manually measured snow depth reveal slight variations in accuracy among the methods, with SRS demonstrating the highest accuracy (correlation coefficient $R^2 = 0.98$), followed closely by remote sensing ($R^2 = 0.96$), whereas the camera-based method (operated for only one winter season) yielded comparatively lower accuracy ($R^2 = 0.93$). Based on our findings, we recommend the installation of an automatic sensor system comprising 2-3 sensors for regular snow monitoring due to its cost-effectiveness, low maintenance requirements, and high reliability. However, manual snow measurements remain essential for ground-truth data and sensor calibration or adjustment. Additionally, remote sensing data can complement sensor-derived estimates after validation with manual or sensor data. The camera-based method presents a valuable supplementary tool, particularly suitable for citizen-science initiatives in areas lacking electrical power or facing logistical challenges for daily monitoring.

---

[1] Ontario Ministry of Environment, Conservation and Parks, Dorset, ON, Canada
[2] Natural Resources Canada, Ottawa, ON, Canada
[3] Ontario Ministry of Natural Resources and Forestry, Peterborough, ON, Canada
[4] Nipissing University, North Bay, ON, Canada
[5] Colorado State University, Fort Collins, CO, USA
Corresponding author: huaxia.yao@ontario.ca

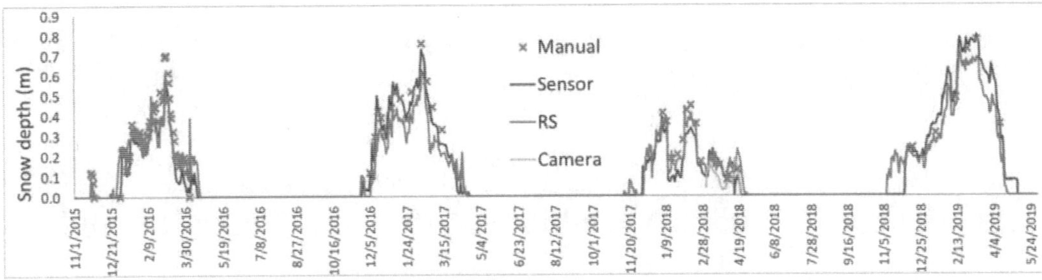

Figure 1. Daily snow depths in four winter seasons of 2015-2019 at Dorset, comparing four methods: sensor, remote-sensing (RS), camera, and manual.

Spatial and Temporal Variability of 2023 Hail Precipitation in Calgary, Alberta, Canada

JULIA M. FINLAYSON[1] AND KRYSTOPHER J. CHUTKO[2]

ABSTRACT

This paper is set to seek out the spatial and temporal distribution of hail in Calgary during the 2023 summer season (June to August). Calgary is located in southern-central Alberta, approximately 80 km east of the Rocky Mountains. It is hypothesized that hail fall will be greatest in western Calgary (i.e., most adjacent to the mountains), though unevenly distributed across the city. Hailstone quantity and diameter data were obtained from acoustic-type disdrometers installed in and around Calgary by the Northern Hail Project team at The University of Western Ontario. Results show that the largest hailstorms in 2023 originated west of the city and moved eastward over the city. Measurements showed that the greatest amount of hail was received in the northwest quadrant of the city and that overall, the northern half received more hail than the southern half. The size of the measured hailstones followed a similar pattern. Temporal analysis showed that hail events in Calgary last approximately 20-30 minutes with the greatest number and largest hailstones occurring in the first half of the storm. These results provide data that may be useful for assessing and mitigating the impact of hailstones on industry and infrastructure in and around Calgary.

[1] College of Agriculture and Bioresources, University of Saskatchewan, Saskatoon, SK, Canada
[2] Department of Geography and Planning, University of Saskatchewan, Saskatoon, SK, Canada
Corresponding author: jfinla23@uwo.ca

Groundwater Loss Continues in the Great Basin of the Western United States, Despite Annual Replenishment by Snow*

DOROTHY K. HALL[1,2], BRYANT D. LOOMIS[3], NICOLO E. DIGIROLAMO[2,4], AND BARTON A. FORMAN[5]

ABSTRACT

An acceleration of groundwater loss in the Great Basin of the western United States has occurred since the southwestern U.S. megadrought began in 1999. This may be seen from changes in the Earth's gravity measured by the Gravity Recovery And Climate Experiment (GRACE) and GRACE-Follow On (FO) satellites. Groundwater is a major component of total water storage (TWS), and snowfall and snowmelt in the Great Basin are the major causes of seasonal fluctuations of TWS. TWS consists of groundwater, surface water, snow and ice, soil moisture and atmospheric water vapor. Surface water, soil moisture and atmospheric water vapor—are not major constituents of TWS in this arid/semi-arid region. Snow and groundwater are the primary components of the TWS in the arid/semi-arid Great Basin. As a snowpack accumulates or melts, water is redistributed causing a rapid regional change in gravity that can be measured from space. From 2002–2023, there is a substantial decline of groundwater in the Great Basin of ~68.7 km^3 which is more than six times the current volume of water in the Lake Mead Reservoir in Arizona/Nevada. Stronger declining trends of groundwater are found in the western part of the Great Basin while weaker declining trends are found in the eastern part (e.g., Great Salt Lake Basin). Despite annual replenishment by snowfall, the downward trend of groundwater depletion persists. Likely causes include declining snow mass, upstream water diversions and increased evaporation/sublimation due to increasing air and surface temperatures. Groundwater depletion is associated with dire consequences for humans and wildlife.

*From: Hall DK, Loomis BD, DiGirolamo NE, Forman BA. 2024. Snowfall replenishes groundwater loss in the Great Basin of the Western United States, but cannot compensate for increasing aridification. *Geophysical Research Letters*, **51:** e2023GL107913. *https://doi.org:10.1029/2023GL107913*

[1] Earth System Science Interdisciplinary Center/University of Maryland, College Park, MD, USA
[2] Cryospheric Sciences Lab/NASA/Goddard Space Flight Center, Greenbelt, MD, USA
[3] Geodesy and Geophysics Lab/NASA/Goddard Space Flight Center, Greenbelt, MD, USA
[4] SSAI, Lanham, MD, USA
[5] Department of Civil and Environmental Engineering, University of Maryland, College Park, MD, USA
Corresponding author: dkhall1@umd.edu

Towards Improving the Canadian Land Data Assimilation System (CaLDAS) for Reliable Snow Products in Support of the Terrestrial Snow Mass Mission (TSMM)

COURTNEY BAYER[1], VINCENT VIONNET[1], BERNARD BILODEAU[1], MARCO CARRERA[1], NICOLAS R. LEROUX[1], AND VINCENT FORTIN[1]

ABSTRACT

The Terrestrial Snow Mass Mission (TSMM) aims to launch a double Ku-band satellite whose measurements will be integrated into modelling systems to improve estimates of snow mass over land. Currently, snow water equivalent (SWE) information in Canada is insufficient (especially at high elevations and high latitudes), but is of great importance for livelihoods, including improving our knowledge of water security. For this mission, data assimilation of synthetic aperture radar (SAR) measurements will occur in the Canadian Land Data Assimilation System (CaLDAS), that relies on ensemble-based assimilation methods. Before data assimilation can occur, it is important that the snow ensemble from the assimilation system is verified for optimal performance. An ensemble of snowpack simulations without data assimilation was run with CaLDAS over the Ottawa River Basin from October 2021-October 2022. This ensemble was found to be under-dispersive and smoothed the SWE and snow depth in complex terrain. To improve the existing ensemble techniques implemented in CaLDAS, the system is re-run multiple times with changes to the precipitation phase partitioning and perturbation methods. Ensemble verification metrics for bias, accuracy, skill and reliability are evaluated for the 2021/2022 winter and the different ensemble runs compared. The results of this evaluation will provide insight into the ideal ensemble configuration, including optimal precipitation phase partitioning and perturbation methods, which will be used during the assimilation of observations retrieved from the TSMM satellite to improve snow estimates.

---

[1] Meteorological Research Division, Environment and Climate Change Canada, Dorval, QC, Canada
Corresponding author: courtney.bayer@ec.gc.ca

Empowering Tomorrow's Climate Science: The University Consortium for the Canadian Terrestrial Snow Mass Mission (TSMM)

ALEXANDRE LANGLOIS[1,2] AND NICOLAS MARCHAND[1,2]

ABSTRACT

The urgent need for precise monitoring of terrestrial snow mass, driven by the accelerating pace of climate change, has prompted the establishment of the Canadian University Consortium for the Terrestrial Snow Mass Mission (TSMM). This initiative, led by snow researchers from 13 Canadian universities, aims to address gaps in our understanding of Earth's water cycle. By leveraging advanced space technology, the consortium focuses on quantifying snow cover on Earth's surface, an area previously overlooked by dedicated satellite missions. The consortium operates under three guiding principles: demonstrating interest, driving innovation, and advocating for a dedicated satellite mission. It actively promotes the TSMM's significance through various research activities and collaborations with Environment and Climate Change Canada, aiming to raise awareness among scientific communities, government bodies, and the public. Innovation is central to its mandate, with a commitment to advancing satellite technology and retrieval algorithms for more accurate snow mass measurements.

Recent advancements, such as improved understanding of microwave responses to snow microstructural properties and radar sensors capable of retrieving snow depth up to 4 meters, have bolstered the feasibility of the TSMM concept. The consortium envisions a future where comprehensive data fills the gaps in Earth's water cycle, facilitated by collaborative efforts among universities and stakeholders. Operating within a structured framework led by the Centre d'applications et recherches en télédétection (CARTEL) at the Université de Sherbrooke, the consortium facilitates knowledge exchange through newsletters and biannual calls, fostering collaboration and alignment of efforts. By engaging with key entities like Environment and Climate Change Canada and the Canadian Space Agency, the consortium communicates the importance of TSMM in addressing global challenges related to snow mass measurement and climate science. This collaborative approach not only enhances our understanding of terrestrial snow mass but also contributes to climate science, water resource management, and disaster preparedness.

[1] Centre d'Applications et de Recherches en Télédétection (CARTEL), Université de Sherbrooke, Sherbrooke, QC, Canada

[2] Groupe de Recherche Interdisciplinaire sur les Milieux Polaires (GRIMP)

Corresponding author : a.langlois@usherbrooke.ca

Estimating Lake-Ice Thickness using Sentinel-3B SRAL: A Case Study on Kluane Lake

JENNIFER FATT[1]

ABSTRACT

Obtaining accurate and consistent estimates for lake-ice thickness in arctic and sub-arctic environments is crucial for our understanding of regional hydrology, ecology, and climate in addition to creating and implementing socioeconomic policies and considerations for surrounding communities. Lake ice plays a large role in the lives and wellbeing of populations around the lake and within the surrounding area, particularly for those who live in remote regions and rely on the lake for a means of transportation on ice roads, recreational use, subsistence hunting and fishing, as well as cultural and religious practices. This research investigates the use of the synthetic aperture radar altimeter (SRAL) instrument aboard Sentinel-3B to estimate lake ice thickness for Kluane Lake in the Yukon through waveform analysis and ground truth measurement comparison. Due to its ability to measure surface heights with high precision and penetrate cloud cover, the Sentinel-3B altimeter offers an opportunity to contribute to filling the knowledge gap created by the decline in *in situ* observations and weather stations. This research aims to contribute to climate and hydrological models and improve the overall understanding of lake-ice dynamics and environmental changes in high latitude regions.

[1] University of Waterloo, Waterloo, ON, Canada
Corresponding author: jcfatt@uwaterloo.ca

Optimized Pixel-Specific Snow Melt Detection Process and Assessment for Calibrated Enhanced-Resolution Brightness Temperatures (CETB)

JOAN M. RAMAGE[1], MAHBOUBEH BOUESHAGH[1], MARY J. BRODZIK[2], AND MOLLY A. HARDMAN[2]

ABSTRACT

Snow melt timing and magnitude are important for hydrology, ecology, energy balance, and human systems. Calibrated Enhanced-Resolution brightness temperatures (CETB) offer an approach to refine melt timing algorithms in remote and high latitude frozen regions because of their sensitivity to liquid water in the snowpack and their ability to detect surface conditions through cloud and darkness. Snow melt is usually identified on the date where both the brightness temperature and the diurnal amplitude variation (DAV), defined as the absolute value of the difference between the successive morning and evening overpasses ($|Tb_n - Tb_{n+1}|$), exceed melt condition thresholds. Legacy thresholds are 246K / ±10 K for Special Sensor Microwave Imager (SSM/I). Twice-daily enhanced-resolution observations are produced at 3.125 km spatial resolution for the 37 GHz channels. In this presentation, we share new work to identify and evaluate pixel-specific thresholds at 37GHz vertically polarized channels based on sites in Alaska, USA. Refined thresholds for Tb and DAV are based on auxiliary statistics and time series analysis. We compare the results to ground based datasets such as snow water equivalent (SWE) depletion records, stream discharge, air temperature and landcover and site characteristics. We compare the outcomes of using legacy Tb and DAV thresholds versus updated site-specific thresholds. Tb thresholds have a range of values across sites and time periods; DAV thresholds seem robust and consistent with the legacy values. These refinements show the promise of optimized thresholds for global snow melt characterization in the absence of *in situ* measurements.

[1] Earth and Environmental Sciences, Lehigh University, Bethlehem, PA, USA
[2] NSIDC/CIRES, University of Colorado at Boulder, Boulder, CO, USA
Corresponding author: ramage@lehigh.edu

Evaluating RADARSAT Constellation Mission Data for the Retrieval of Lake Ice Melt Events over Sub-Arctic Regions in Canada

SABA DEZYANI[1] AND GRANT E. GUNN[1]

ABSTRACT

Lake ice is a significant component of the physical landscape at northern latitudes and plays a crucial role in regulating weather and climate, as well as facilitating winter transportation between northern towns. The sensitivity of lake ice to air temperature makes it an effective indicator of climate change and it is crucial to closely observe variations in lake ice to gain insight into the potential impact that lakes might have on regional weather patterns and environmental conditions. Satellite data products offer the advantage of gathering data consistently over a region of interest, which is particularly useful in remote locations. This study aims to determine melt timings of lake ice in two selected lakes in Canada's sub-Arctic regions where ground observation is limited. Specifically, we focus on Kluane Lake in the Yukon Territory, renowned for its vast area and depth, and Landing Lake in the Northwest Territories, notable for its shallow depth. RADARSAT Constellation Mission (RCM) acquisitions, especially compact polarimetry data with the innovative circular polarization method will be used to monitor lake ice changes over the lake ice break-up period. The study will employ changes in microwave backscatter intensity and a first difference method to detect changes between acquisitions and extract ice melt events from RCM data. To validate RCM's accuracy in extracted melt dates, results will be compared with thermodynamic lake ice models and meteorological data from weather stations. Through this study, we anticipate achieving a nuanced understanding of the strengths and potential limitations of the compact polarimetry that is collected by RCM's standard coverage to capture lake ice phenology over sub-Arctic lakes in Canada with different morphometric characteristics.

[1] Department of Geography and Environmental Management, University of Waterloo, Waterloo, Ontario
Corresponding author: sdezyani@uwaterloo.ca

Improving Passive Microwave Estimation of Snow-Covered Land Surface Temperature

QUINGUAN LI[1] AND RICHARD KELLY[1]

ABSTRACT

Land Surface Temperature (LST) plays a pivotal role in retrieving snow depth (SD) and snow water equivalent (SWE). It influences ground emissions and impacts the metamorphism of snowpacks. The size of snow grains, which is crucial for SD and SWE retrieval, depends on temperature fluctuations. Given these fluctuations, relying solely on satellite or in-situ observations for SD and SWE operational retrieval proves unreliable. The snowpack significantly influences ground emissions. In conjunction, physical temperature challenges the accurate estimation of LST in snow-covered regions using passive microwave techniques. Although algorithms for LST estimation based on passive microwave observations exist, they are not optimally suited for snow-covered conditions.

Passive microwave techniques offer an independent LST estimation applicable globally under all-weather conditions, complementing weather station, near-infrared, and model simulation data. These techniques also support the operational retrieval of SD and SWE using the same sensor, such as AMSR2. Despite their relatively lower accuracy in snow-covered conditions compared to optical and near-infrared observation methods, refining passive microwave-based LST estimation remains essential. This study delves into the uncertainties inherent in the original LST estimator within the SMSA model. It proposes an ANN model tailored for snow-covered condition LST estimation, trained using weather station data, ERA5 data, and model simulations.

[1] University of Waterloo, Waterloo, ON, Canada
Corresponding author: liqinghuan@gmail.com

Role of Snow in Reducing Underwater Photosynthetically Active Radiation in Ice-Covered Lakes

MADELINE E. MYERS[1], PETER T. DORAN[2], AND KRISTA F. MYERS[2]

ABSTRACT

Under climate warming conditions, snow is expected to increase in polar regions. It is well-known that snow and ice affect the underwater light climate and primary productivity, but the relationship between snow and underwater photosynthetically active radiation (UW PAR) in perennially ice-covered lakes is less understood. Here, I leverage the long-term monitoring of surface PAR, UW PAR, and snow cover to quantify the relationship between snow and underwater light climate in three major lakes of Taylor Valley, Antarctica. We find snow accumulation of only 0.5 mm w.e. on the entire lake surface can suppress UW PAR by up to 40%. Following a ~7 mm w.e. snowfall event in 2016, UW PAR was suppressed below prior values for as much as 22 days at a lake with thicker ice and 10 days at lakes with thinner ice. When simulating UW PAR reduction for snow thicknesses of 0.5 to 2 cm at different times of year and different lakes, we found that events during shoulder seasons are more likely to slow primary productivity. Furthermore, the lakeshore camera record revealed a tripling of days per year with snow cover beginning in 2012, most notably in the shoulder seasons. We see a trend in increasing snow persistence during the Fall over the 10-year camera record, although the relationship is not statistically significant. Empirical relationships rule out increased volume and frequency of snow events as the cause. If snow cover continues to increase with climate warming, the seasonality of snow events may play a greater role than the magnitude at reducing UW PAR and associated primary productivity.

[1] Department of Geography and Planning, Queen's University, Kingston, ON, Canada
[2] Department of Geology and Geophysics, Louisiana State University, Baton Rouge, LA, USA
Corresponding author: 19mem7@queensu.ca

Snowfall from Lake Water Pressure: Results from Four Lakes in Humid Continental Climate

SAMUEL E. TUTTLE[1], COLIN M. BEIER[2], JAMES S. MILLS[3], GRANT E. GUNN[4], IAN HALM[5], ANA M. MORALES-WILLIAMS[6], AND HAMISH D. PRITCHARD[7]

ABSTRACT

Snowfall is typically measured at the point scale and is vulnerable to several sources of measurement error and spatial representativeness uncertainty. However, accurate snowfall observations are critical inputs for snowpack models. Here, we use a recently published method (Pritchard et al., 2021, *J. Hydrometeorology*) to infer the mass of snowfall falling onto a given lake from the increase in water pressure measured in the lake. This method has the potential to increase the footprint over which snowfall is typically measured by multiple orders of magnitude, which has the added benefit of reducing the spatial scale discrepancy between ground-based observations and the resolution of weather and climate models. Because the entire lake surface acts as a giant snow pillow, measurement errors that can arise in conventional, point-scale snowfall measurements are reduced. We expand on previous results evaluating the viability of the method in a lake in the Adirondack Mountains of New York, USA, to include three additional lakes in the Northeastern USA and southern Canada: Sunfish Lake (Waterloo, ON, Canada), Mirror Lake (Woodstock, NH, USA), and Stannard Pond (Stannard, VT, USA). These sites help to evaluate whether the method can be applied in regions with temperate winter climates. Lake-estimated snowfall is compared to nearby snow and meteorological observations in order to evaluate whether the method can be applied in regions with temperate winter climates. Agreement between the lake-estimated snowfall and a nearby snowfall observations vary by site, but can reach root mean squared error (RMSE) values of approximately 2 mm and R^2 values above 0.95, supporting the viability and accuracy of the lake pressure-snowfall method.

---

[1] Department of Earth and Environmental Sciences, Syracuse University, Syracuse, NY, USA

[2] State University of New York College of Environmental Science and Forestry, Syracuse, NY, USA

[3] State University of New York College of Environmental Science and Forestry Newcomb Campus, Newcomb, NY, USA

[4] Department of Geography and Environmental Management, University of Waterloo, Waterloo, ON, Canada

[5] U.S. Forest Service, Northern Research Station, Hubbard Brook Experimental Forest, North Woodstock, NH, USA

[6] Rubenstein School of Environment and Natural Resources, The University of Vermont, Burlington, VT, USA

[7] British Antarctic Survey, High Cross, Cambridge, United Kingdom

Corresponding author: setuttle@syr.edu

Oral Session #4

Session Chairs: Peter Toose & Tyler Herrington

New Estimates of Decreasing September-March Snow Cover from Multi-Dataset Analysis

ALEKSANDRA ELIAS CHEREQUE[1], PAUL KUSHNER[1], CHRIS DERKSEN[2], AND LAWRENCE MUDRYK[3]

ABSTRACT

Large-scale studies of snow cover variability and change have primarily been performed using data derived from satellite observations. However, these datasets can be limited by discontinuities that are present in the observational record; changes in instrument resolution, retrieval algorithms, and different orbits of different satellites all impact a record's temporal stability. This fact has led to controversy about the popular NOAA Climate Data Record of Snow Cover Extent (NOAA CDR), which reports sizeable increases in snow cover extent (SCE) over the 1980-2020 period that are in opposition to trends derived from station observations. We contribute a new line of evidence by producing and analyzing a range of snow cover extent datasets based on modeled snow water equivalent (SWE). These datasets are spatially complete, and they perform well against in-situ SWE observations, so they are suitable for hemisphere-scale study. We find that no fixed threshold to convert SWE to SCE can reproduce the trends in the NOAA CDR for any month. Nonetheless, all the SCE records (expressed as anomalies from a running mean rather than the raw values) give similar rankings of years by SCE. Further investigations reveal results that are robust across all the months of study (September-March): to reproduce the trends of the NOAA CDR with the historical snow cover data, we must systematically reduce the SWE threshold that defines snow cover. This result is consistent with an observing system that has undergone improvements in its ability to discern snow cover over time, a mechanism that can explain the NOAA trend. We use the last ten years of data from the NOAA CDR to calibrate and produce a merged SCE dataset for 1980-2020. Monthly trend estimates can be used in future applications to validate climate model output.

---

[1] University of Toronto, Department of Physics, Toronto, ON, Canada
[2] Environment and Climate Change Canada, Climate Research Division, Toronto, ON, Canada
Corresponding author: Aleksandra.eliaschereque@mail.utoronto.ca

Table 1. Snow cover extent targets for calibration and trends for September to March based on the forty-year period from August 1980 to July 2020. Trend significance ($p < 0.05$) is indicated by bold text.

	North America		Eurasia		Northern Hemisphere	
Month	Total Area (million km^2)	Trend (million km^2/dec)	Total Area (million km^2)	Trend (million km^2/dec)	Total Area (million km^2)	Trend (million km^2/dec)
September	1.77	**-0.15±0.01**	1.73	-0.18±0.02	3.51	**-0.33±0.02**
October	6.28	**-0.21±0.02**	11.38	**-0.31±0.03**	17.66	**-0.52±0.04**
November	11.40	**-0.25±0.02**	20.45	**-0.55±0.03**	31.84	**-0.8±0.04**
December	13.44	-0.11±0.01	24.80	**-0.3±0.03**	38.24	**-0.41±0.03**
January	13.90	-0.06±0.01	26.26	-0.17±0.03	40.16	-0.23±0.03
February	13.66	0.01±0.01	25.23	**-0.35±0.03**	38.89	-0.35±0.04
March	12.51	-0.04±0.02	21.40	**-0.62±0.04**	33.90	**-0.65±0.04**

Large Scale Tundra Snow Density Estimates from Passive Microwave Remote Sensing

JEFFREY J. WELCH[1]

EXTENDED ABSTRACT

Snow density data are useful for a number of applications but typically they are few and far between. Much of the current understanding of snow density is limited to distributed *in situ* samples, which are laboursome to collect and the point-based measurements they produce lead to uncertainties when extrapolated for use more broadly. In a remote environment, like the Canadian Arctic, it is not feasible to collect extensive *in situ* samples to characterize snow density on large-scales and, instead, it can be better estimated with models. However, current snow density products from models cannot adequately predict tundra snow density (Cao *et al.*, 2020; Domine *et al.*, 2019). This research builds upon a previous study (Welch, 2023) to estimate snow density using satellite passive microwave remote sensing in the tundra environment.

Tundra snow is composed of a characteristic two-layer structure, consisting of a lower depth hoar layer overlain by a denser wind slab layer (Benson & Sturm, 1993), so a two-layer electromagnetic snowpack model is used to estimate microwave emissions in the remote sensing retrieval algorithm. The two-layer model introduced an issue of equifinality because multiple layer density combinations could produce the same synthetic emissions resulting in a system with no global minima. The proposed algorithm addresses the issue of equifinality by solving for all plausible layer density combinations and is hence named the Multiple Answers for Inversion (MAIN) algorithm. The MAIN algorithm uses boundary conditions imposed on the snowpack model to constrain snow density estimates for each snow layer to a plausible range.

There are two boundary conditions for the snow density parameters in the MAIN algorithm. The lower boundary condition for the model is related to strong environmental controls in the tundra that produce a dense wind slab layer on top of a lighter depth hoar layer. Therefore, it is assumed that the wind slab density should always be greater than the depth hoar layer and the lower boundary condition for the model exists where the two snow layers have equal densities. The upper boundary condition is related to the effect of snow density on microwave emissions in the Dense Media Radiative Transfer model (DMRT) for multilayered snowpacks, where the relationships between snow density and microwave scattering and absorption are consistent between 150-450 kg m^{-3} (Picard *et al.*, 2013). Thus, the domain of each layer snow density is constrained to 150-450 kg m^{-3} and the upper boundary condition for the model exists where either layer was at the edge of its domain.

With the two established boundary conditions, the MAIN algorithm can solve for all plausible layer density combinations for a given scene. The cost surface (defined by the difference between observed and estimated microwave emissions) is characterized by the local minima forming a valley transecting the parameter space (Fig. 1). Thus, the solutions at the boundary conditions (red points in Fig.1) can be connected to form a set of the DMRT-plausible layer density combinations (red line

[1] University of Waterloo, Waterloo, ON, Canada
Corresponding author: jjwelch@uwaterloo.ca

in Fig. 1). The DMRT-plausible estimates can then be converted to bulk estimates of snow density and preliminary results agree well with *in situ* reference samples (see Fig 2.).

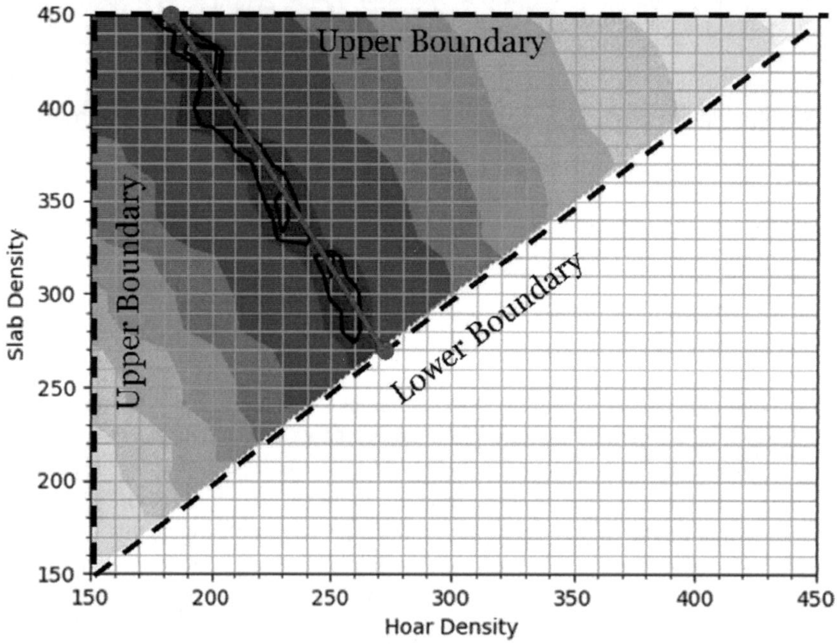

Figure 1. Model parameter space of potential wind slab and depth hoar density combinations, darker (lighter) shades of blue represent lower (higher) cost (i.e. difference between simulated and observed emissions). Red points represent solutions at boundary conditions (dashed lines) and the set of DMRT-plausible combinations between them along the red line.

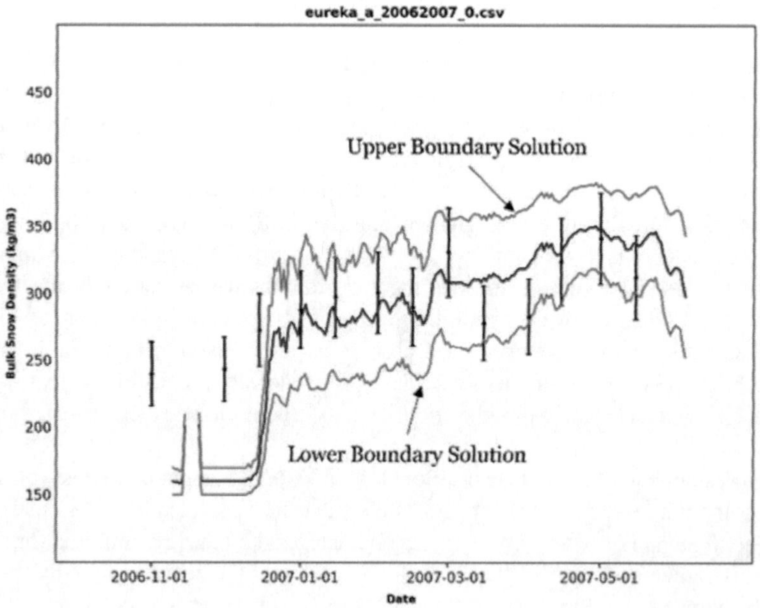

Figure 2. MAIN algorithm estimates of bulk snow density, for Eureka 2006-07, at the model boundary conditions (grey lines) and the final estimate set halfway between (blue line), plotted with *in situ* density samples (points, with 10% error bars) from the Canadian Historical Snow Water Equivalent dataset (Vionnet *et al.*, 2021).

REFERENCES

Benson CS, Sturm M. 1993. Structure and wind transport of seasonal snow on the Arctic slope of Alaska. *Annals of Glaciology*, **18:** 261–267. *https://doi.org/10.3189/s0260305500011629*

Cao B, Gruber S, Zheng D, Li X. 2020. The ERA5-Land soil temperature bias in permafrost regions. *Cryosphere*, **14:** 2581–2595. *https://doi.org/10.5194/tc-14-2581-2020*

Domine F, Picard G, Morin S, Barrere M, Madore JB, Langlois A. 2019. Major issues in simulating some Arctic snowpack properties using current detailed snow physics models: consequences for the thermal regime and water budget of permafrost. *Journal of Advances in Modeling Earth Systems*, **11**: 34–44. *https://doi.org/10.1029/2018MS001445*

Picard G, Brucker L, Roy A, Dupont F, Fily M, Royer A, Harlow C. 2013. Simulation of the microwave emission of multi-layered snowpacks using the Dense Media Radiative transfer theory: the DMRT-ML model. *Geoscientific Model Development*, **6**: 1061–1078. *https://doi.org/10.5194/gmd-6-1061-2013*

Vionnet V, Mortimer C, Brady M, Arnal L, Brown R. 2021. Canadian historical Snow Water Equivalent dataset (CanSWE, 1928–2020). *Earth System Science Data*, **13**: 4603–4619. *https://doi.org/10.5194/essd-13-4603-2021*

Welch JJ. 2023. Monitoring Seasonal Snow Density from Satellite Based Passive Microwave Remote Sensing and Automatic Weather Stations. University of Waterloo.

Spatially Distributed Modelling of Lake Ice Thickness Trends and Distribution in the North Slave Region, NWT

GIFTY ATTIAH[1], HOMA KHEYROLLAH POUR[1], AND K. ANDREA SCOTT[2]

ABSTRACT

Lake ice thickness (LIT) is a critical parameter for understanding the overall stability of lake ice, which directly impacts the safety of the construction of ice roads construction and winter travel over frozen lakes in northern communities. Optical and thermal sensors, while they have become invaluable tools for monitoring lake ice cover and phenology - such as freeze-up, break-up, and duration – cannot measure LIT directly due to its subsurface nature. This study employs a multimethod approach, integrating remote sensing data with a spatially distributed thermodynamic lake ice model to simulate and map ice thickness on sub-arctic lakes.

We modelled the spatial variability of LIT across 535 predominantly small to medium-sized lakes in the North Slave Region of the Northwest Territories (NWT) from 1984 to 2022. To effectively simulate small lakes, lake surface temperature data were extracted from the Landsat satellite series, generating the North Slave LST dataset, which served as model inputs alongside ERA5 reanalysis data. LIT simulations were based on an unsteady heat equation balance, with model variables including wind speed (m s-1), mean air temperature (°C), relative humidity (%), snow depth (m) and cloud cover (0-1) extracted from these gridded datasets. The model outputs include daily lake ice thickness on a 50m grid over lakes, as well as annual freeze-up and break-up dates. Results showed a root mean square deviation of lake ice thickness from 3 cm to 7 cm compared to in-situ data. Further analysis indicated decreasing trends in ice thickness (-0.26 cm/year to -0.10 cm/year) and ice cover duration (-0.40 day/year to -0.15 day/year). Additionally, later freeze-up (0.10 day/year to 0.30 day/year) and earlier break-up (-0.16 day/year to -0.21 day/year) was observed. This research presents a replicable method for estimating ice thickness on small to medium-sized lakes, offering valuable insight for safety and planning purposes.

---

[1] Department of Geography and Environmental Studies, Wilfrid Laurier University, Waterloo, ON, Canada.

[2] Department of Systems Design Engineering, University of Waterloo, Waterloo, ON, Canada

Corresponding author: gattiah@wlu.ca

Leveraging Satellite Imagery and Machine Learning to Track Equilibrium Line Altitude and Surface Mass Balance Trends in Auyuittuq National Park, Canadian Arctic

WILSON (WAI YIN) CHEUNG[1] AND LAURA THOMSON[1]

ABSTRACT

The equilibrium line altitude (ELA) is the elevation on a glacier where annual snow accumulation equals annual snow and ice melt, delineating the upper accumulation zone from the lower ablation zone. The precise ELA location depends on climatic factors like temperature, precipitation, and solar radiation. Tracking ELA changes provides insights into glacier surface mass balance (SMB) variability. As climate warms, a rising ELA indicates negative SMB, with more area in the ablation zone experiencing net ice loss. Conversely, a lower ELA suggests positive SMB and accumulation zone expansion.

The Canadian Arctic Archipelago hosts one-fifth of the global glacierized terrain, yet knowledge of spatial glacier response and sea-level rise contributions in this region remains uncertain. While the Canadian Arctic has the highest density of long-term glacier SMB records, most data is concentrated above 74°N latitude. A persistent data gap exists across the glaciers of Baffin and Bylot Islands (BBI), hampering comprehensive understanding of temporal and spatial SMB variations.

This study develops a semi-automated methodology using machine learning and satellite imagery to classify snow/ice and derive annual ELAs across the Auyuittuq National Park in Nunavut, Canada. By applying the ELA-mass balance gradient approach, we reconstruct annual SMB estimates for these glaciers from 2015-2023. This research sheds light on the implications of ice mass loss within the park, crucial for discerning ongoing climate dynamics in this data-sparse region.

[1] Ice Climate and Environment Lab (ICElab), Queen's University, Department of Geography and Planning, Kingston, ON, Canada

Corresponding author: w.cheung@queensu.ca

Oral Session #5

Session Chairs: Jennifer Jacobs & Alicia Pouw

Status and Progress of the Terrestrial Snow Mass Mission

BENOIT MONTPETIT[1,2], CHRIS DERKSEN[1], VINCENT VIONNET[1], MARCO CARRERA[1], JULIEN MELOCHE[1], NICOLAS LEROUX[1], MICHAEL BRADY[1], PETER TOOSE[1], VINCENT FORTIN[1], COURTNEY BAYER[1], STÉPHANE BÉLAIR[1], RICHARD KELLY[3], ALEXANDRE LANGLOIS[2,4], PAUL SIQUEIRA[5], AARON THOMPSON[3], NICOLAS ALLET[2,4], JEAN BERGERON[6], PATRICK PLOURDE[6], AND RALPH GIRARD[6]

ABSTRACT

Within the complete water cycle, snow is its only component that does not currently have a dedicated earth observation mission. Snow impacts many sectors like the well-being of communities, economy, and sustains ecosystems. Snow also contributes to many hazards (floods, droughts, and avalanches). The current lack of information on snow water equivalent (SWE), limits the hydrological, climate, and weather services provided by Environment and Climate Change Canada (ECCC). To address this, ECCC and the Canadian Space Agency (CSA) are working to advance the scientific and technical readiness for a synthetic aperture radar (SAR) mission presently named the 'Terrestrial Snow Mass Mission' – TSMM. An observing concept capable of providing dual-polarization (VV/VH), moderate resolution (500 m), wide swath (~250 km), and high duty cycle (~25% SAR-on time) at two frequencies (13.5; 17.25 GHz) is under development. Ku-Band backscatter is proven to be sensitive to SWE through volume scattering by the snow microstructure.

In this presentation, we will review recent progress at ECCC (supported by the mission science team and the international snow community). Key areas of ongoing development include:

(1) Implementation of computationally efficient SWE retrieval techniques, including forward modeling parameterization and reducing snowpack layering to radar relevant layers from land surface models.

(2) Incorporation of land surface model SWE estimates to infill gaps for which no radar-derived SWE information will be retrievable.

(3) Development of the capability for direct assimilation of Ku-band backscatter into environmental prediction systems at ECCC.

(4) Further validate the technical concept of TSMM via field campaigns.

Testbed experiments in which snow physical modeling, SWE retrievals, and data assimilation are currently under development. These experiments are facilitated by the TSMM simulator, will incorporate outputs from the ECCC land surface prediction system, and are supported by airborne and ground-based Ku-band radar measurements from national and international academic partners.

---

[1] Environment and Climate Change Canada (ECCC)

[2] Centre d'Applications et de Recherches en Télédétection (CARTEL), Université de Sherbrooke, Sherbrooke, QC, Canada

[3] University of Waterloo, Waterloo, ON, Canada

[4] Groupe de Recherche Interdisciplinaire sur les Milieux Polaires (GRIMP)

[5] University of Massachusetts Amherst, Amherst, MA, USA

[6] Canadian Space Agency (CSA)

Corresponding author: benoit.montpetit@ec.gc.ca

SNOWWI - A Three-Frequency SAR and InSAR for Measuring Snow Characteristics: Early Results from the 2023-24 Season Grand Mesa Campaign

PAUL R. SIQUEIRA[1], J. MAX ADAM[1], MARC CLOSA TARRES[1], ERIC SUTHERLAND[1], JOSEPH MALOYAN[1], TAKUYA SEAVER[1], HANS-PETER MARSHALL[2], ELIAS DEEB[3], AND KELLY ELDER[4]

ABSTRACT

Supported with funds from NASA's Earth Science Technology Office (ESTO) Instrument Incubator Program (IIP), the University of Massachusetts (UMass) has been developing a three-frequency (C-band @ 5.39 GHz, Ku-band low 13.64 GHz, and Ku-band high 17.24 GHz) dual-polarized SAR interferometer called SNOWWI (Snow Water Equivalent Wide-Swath Interferometer). Although it is still in its early stages of development, the SNOWWI instrument is expected to have a signal to noise ratio (or noise equivalent sigma zero; NES0) that would be sufficient for estimates of Snow Water Equivalent over a wide swath (NES0 less than -24 dB m^2/m^2 at 10 m ground resolution over a swath width that is greater than 4 km). During the weeks of January 28 – February 4, and March 24 – 29, 2024, and with a concerted ground validation coordinated by Boise State University, CRREL, and the US Forest Service, the SNOWWI instrument was mounted on a Cessna 208 and collected more than 20 TB of data that is now being processed at UMass to demonstrate some of the capability of the instrument.

In this talk, we will be demonstrating some of the basic capability of SNOWWI, showing examples of its data collection ability in terms of achieving its anticipated native resolution of better than 2 m (80 MHz bandwidth), and its abilities in polarimetry (VV- and VH-polarization) and interferometry. A simple example of some of these early results is shown in Figure 1 below, where it can be seen that the slant range swath width extends from a nadir return at 2400 m to more than 8000 m from the platform to the ground (a 5.6 km swath).

[1] University of Massachusetts, Amherst, MA, USA
[2] Boise State University, Boise, ID, USA
[3] US Army Corps of Engineers Cold Regions Research and Engineering Laboratory, Hanover, NH, USA
[4] US Forest Service, Colorado Springs, CO, USA
Corresponding author: siqueira@umass.edu

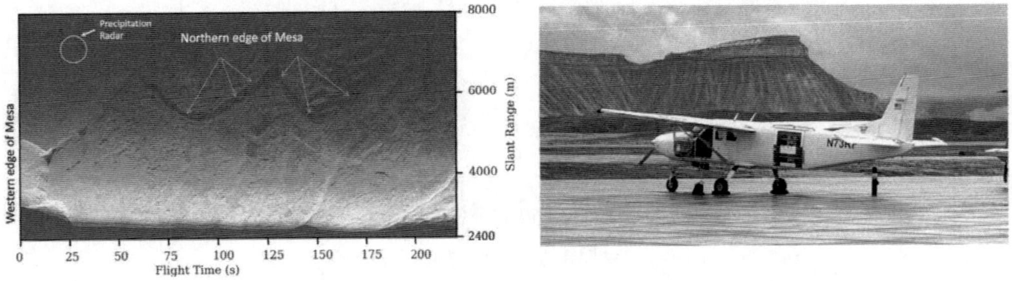

Figure 1. A Ku-band high (17.24 GHz) SNOWWI image of the Grand Mesa region in Western Colorado (left) and an image of the Cessna 208 platform that was used to deploy the instrument.

A New *in situ* Snow Water Equivalent Dataset for the Northern Hemisphere

COLLEEN MORTIMER[1] AND VINCENT VIONNET[2]

ABSTRACT

In situ manual snow water equivalent (SWE) measurements have been conducted for nearly a century yet there is no standard global database to archive these observations, nor is there a standard approach to measuring SWE. It is under this fragmented landscape that we present a new *in situ* snow dataset for the Northern Hemisphere covering the modern satellite era (1979-present). The dataset, to be publicly available on Zenodo as a single NetCDF file, combines snow water equivalent measurements from manual snow surveys, airborne passive gamma radiation SWE, snow pillows and passive gamma radiation sensors covering North America, Russia and Finland (Figure 1).

SWE is the primary variable of interest; SD is included when available and derived bulk snow density is calculated from SD and SWE. A basic quality control, consisting of range thresholding and automated outlier detection is applied (Vionnet et al. 2021). The final dataset, which expands and builds upon the pan-Canadian CanSWE dataset, represents several years' effort to identify, obtain and compile various sources of SWE data. We will provide an overview of this process, highlighting specific challenges involving data access, inadequate metadata, and archived data quality. Summary characteristics of the final public dataset will also be presented.

In situ SWE data, such as those included in our dataset, are necessary for a range of applications from quantifying changes in snow water storage, evaluating hydrological and land surface models and gridded SWE products from satellite observations and climate reanalyses, and to understand their uncertainties. Beyond the primary SWE variable, paired SWE and SD measurements are needed to train and validate snow density models. *In situ* snow density information is also used in snow model verification and to paramaterize snow density in satellite-based SWE retrieval algorithms. Current and future applications of the dataset will be discussed. We hope that publicly sharing these data in an easily accessible format will facilitate additional improvements in snow estimation across a range of applications.

---

[1] Climate Research Division, Environment and Climate Change Canada, Toronto, ON, Canada
[2] Meteorological Research Division, Environment and Climate Change Canada, Dorval, QC, Canada
Corresponding author: colleen.mortimer@ec.gc.ca

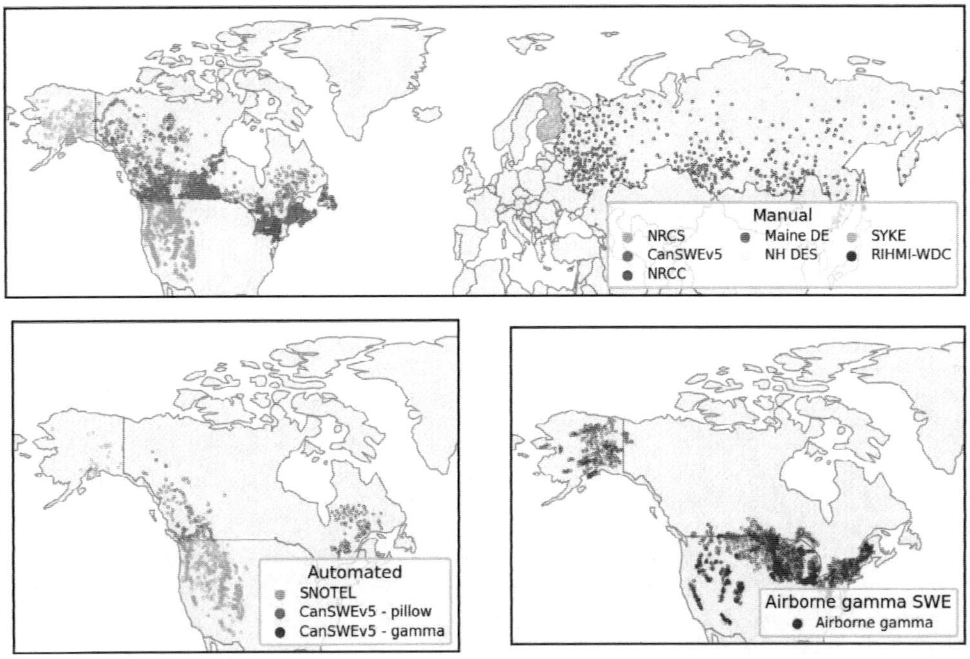

Figure 1. Spatial distribution of manual snow survey, NOAA- NOHRSC airborne gamma SWE, and automated sites included in the Northern Hemisphere in situ SWE dataset. For display purposes, sites are separated by measurement type and colours indicate the data provider. Maine DE - Maine Geological Survey: Maine Cooperative Snow Survey Program, NRCS - U. S. Department of Agriculture Natural Resources Conservation Service, NRCC – Northeast Regional Climate Center, NH DES - New Hampshire Department of Environmental Services – Dams, SYKE – Finish Environmental Institute, RIHMI-WDC - All-Russia Research Institute of Hydrometeorological Information – World Data Center.

Oral Session #6

Session Chairs: Samuel Tuttle & Megan Verfaillie

Climate Scenarios of Snow Water Equivalent Indices for Southern Québec

ÉMILIE BRESSON[1], PASCAL BOURGAULT[1], AND ÉRIC DUPUIS[1]

ABSTRACT

In boreal regions, several economic sectors need snow cover information to plan their winter activities and snowpack melt can also play a crucial role in spring floods. Easily accessible ensemble of climate scenarios of snow cover is in demand among stakeholders and decision-makers. To the best of our knowledge, no such database is publicly available today. The present project aimed to produce such database over Southern Quebec, using snow water equivalent (SWE) simulated by regional climate models and a simple bias adjustment method.

We began by compiling a list of SWE and snow cover indices (e.g., the start, end and duration of continuous snow cover) that are useful for a wide range of sectors and validating it with the target audience. We then retrieved the SWE data produced by regional climate models over a domain including Quebec, where these had been archived. We selected ERA5-Land's SWE as our reference product.

Many models have a bias towards a shorter snow cover duration and the absence of snow cannot be corrected with typical multiplicative adjustments. We thus restricted our ensemble (26 simulations) to a subset (10 simulations) that showed only a small snow cover duration bias. Although this subset is small, it allows model and emission scenario uncertainties to be considered. The tails in end of snow season were also extended with an exponential decay, adding small amounts of snow that help our multiplicative quantile mapping better adjust the seasonality. Based on these bias-adjusted time series, we calculated the annual indices for historical and future periods.

Results showed a decrease of snow cover more pronounced in the south of Québec and in the St. Lawrence valley, than in the rest of the domain; the further North you go, and on certain high mountains, natural snow is slightly better protected from a decline.

[1] Ouranos, Montreal, QC, Canada
Corresponding author: bresson.emilie@ouranos.ca

Simulations of Snow Internal Properties in Forested Environments

NICOLAS R. LEROUX[1], VINCENT VIONNET[1], AND GIULIA MAZZOTTI[2]

ABSTRACT

The boreal forest covers 60% of Canada and is an important ecosystem for carbon capture, climate and hydrology regulation, wildlife refuge, and the Canadian economy. During the winter, processes such as snow canopy interception and unloading affect the evolution of the snowpack and its internal properties (density, microstructure, specific surface area). As measurements of snow in forests are sparse, accurately modelling snow accumulation and melting in forests is essential. The current land surface model used in support of environmental forecasting at Environment and Climate Change Canada (ECCC) is limited to a single-layer snowpack model with incomplete representation of canopy processes. Better representing the internal properties of the snowpack is not only important for remote sensing applications but can also help researchers to better understand wildlife ecology or underlying soil hydrological and thermal regimes. A new version of the Soil, Vegetation, and Snow land surface scheme (SVS-2) that includes the detailed snowpack model Crocus and snow canopy processes is presented. The model was evaluated at Snow Crested Butte, an alpine site in Colorado, USA. Snow depth and snow water equivalent were measured below the forest and in an adjacent open area (clearing). Additional measurements of vertical snowpack stratigraphy (snow grain type) and density were available in the open and forest sites. The model represented the bulk snowpacks properties well (snow depth and snow water equivalent) in both the open and forest sites. Using a temperature dependent parameterization of snow interception in the canopy showed improved representation of intercepted snow at the maritime site than a fix value of interception efficiency. The layered density was usually underestimated in the middle of the winter, particularly at the top of the snowpack, but it was slightly overestimated during the melt. The model better represented the snow grain type in the forest as the simulated snowpack in the open site was warmer than in the observations, resulting in an overestimation of melt-freeze layers. Because snow microstructure properties are important inputs for estimating snow backscatter with radiative transfer models, this new version of the land surface scheme at ECCC will provide the necessary inputs for snow water equivalent retrieval algorithms for the Canadian Terrestrial Snow Mass Mission.

---

[1] Meteorological Research Division, Environment and Climate Change Canada, QC, Canada

[2] Univ. Grenoble Alpes, Université de Toulouse, Météo-France, CNRS, CNRM, Centre d'Études de la Neige, Grenoble, France

Corresponding author : nicolas.leroux@ec.gc.ca

A New Snow Observing Strategy in Support of Hydrological Science and Applications

CARRIE M. VUYOVICH[1], MELISSA L. WRZESIEN[1,2], SUJAY V. KUMAR[1], ETHAN GUTMANN[3], KWO-SEN KUO[1,4], PAUL GROGAN[5], BATUHAN OSMANOGLU[1], MARK CARROLL[1], MICHAEL BAUER[4], DAI-HAI TON THAT[4], NIKLAS GRIESSBAUM[4], BOB ROSENBERG[1,6], JOSUE TAPIA[5], AND HADIS BANAFSHEH[5]

ABSTRACT

Snow accumulation is a seasonally evolving process that results in a reflective, insulating cover over the Earth's land mass, provides water supply to billions of people and supports numerous ecosystems. Snow melt also contributes to short-term and long-term disasters. Snow is a critical storage component of the global water cycle, yet no satellite currently provides global snow water equivalent (SWE) data, the essential information to address hydrologic science questions, at the frequency, resolution and accuracy needed. While snow contributes water resources to a large portion of the Earth's terrestrial area, its coverage and role evolves throughout the season, affecting different regions, elevations and latitudes at different times of the year. The data needs also shift throughout the year. Seasonal snow is an ideal candidate for an optimized observational strategy that leverages existing sensors and focuses future mission concepts on monitoring the most critical areas to provide cost-effective and robust information. In this project we merge multiple technologies to develop a hypothetical experiment and demonstrate a potential snow observing strategy that utilizes multiple diverse data to improve basin-wide SWE and streamflow forecasts. We developed metrics to trigger an additional and/or taskable observations response when observations meet a certain threshold in comparison to the climatology. We assess the value of new potential sensors, such as from commercial smallsats to fill observing gaps and provide higher frequency observation during critical time periods. We also evaluate the potential for focusing higher density observations in regions where concerns for flood, drought or wildfires will benefit from early warning. These dynamic observations could inform use of ground, UAS or airborne observations in regions showing snow volumes outside the normal range or experiencing unexpected snowpack conditions.

[1] NASA Goddard Space Flight Center, Greenbelt, MD, USA
[2] Earth System Science Interdisciplinary Center, University of Maryland, College Park, MD, USA
[3] Research Applications Lab, National Center for Atmospheric Research, Boulder, CO, USA
[4] Bayesics, LLC., Bowie, MD, USA
[5] School of Computing and Augmented Intelligence, Arizona State University, Phoenix, AZ, USA
[6] Science Application Intl Corp, Greenbelt, MD, USA

Snow Distribution and Permafrost Degradation at an Ice-Wedge Polygon Site in the Western Canadian Arctic

ROBIN THORNE[1], BRANDEN WALKER[1], ALEX FOGAL[1],
ROSY TUTTON[2], AND PHILIP MARSH[1]

ABSTRACT

Ice-rich permafrost regions contain ice-wedge polygon terrain that form when frozen ground cracks due to thermal contraction. Ice-wedge polygons occurring in low-lying areas adjacent to thermokarst lakes, which are in abundance across large regions of the Arctic, can create rapid, or catastrophic drainage, due to warming of the permafrost. These drainage events are increasing as of recent, for reasons unknown, and can create extreme floods that are a risk to people and infrastructure located downstream, and the destruction of fish habitat. Ice-wedge polygons can be high- or low-centered where the troughs in-between provide favourable locations for infiltration, water storage and flow. Changes in snow patterns and the ground thermal regime leads to ice-wedge degradation thereby creating deep ponded troughs and the destruction of low-centered polygon rims to create high-centered polygons, altering water and energy balances in the polygonal tundra terrain. To better understand these changes, we use a combination of field observations and drones to monitor changes in the spatial snow distribution and melt and permafrost degradation, at an ice-wedge polygon site near the Trail Valley Creek research observatory, north of Inuvik, NT. Our analysis at an ultra-fine resolution allows us to examine the impact of snow depths on the ground thermal regime. Our findings show that the ice-wedge polygon site was a net sink of snow compared to the surrounding basin, with more snow found in the troughs. The highly variable spatial snow distribution had an impact on the variability of active layer depths, where shallow snow cover initiates an earlier thaw of the active layer. This study found that changing the snow depth at ice-wedge polygon sites is likely as important as warming winter air temperature in controlling ice-wedge temperature, cracking, and drainage.

---

[1] Wilfrid Laurier University, Waterloo, ON, Canada
[2] McMaster University, Whitehorse, YT, Canada
Corresponding author: rthorne@wlu.ca

Oral Session #7

Session Chairs: George Riggs & Jack Dechow

High-Resolution Drone Mapping of Snow Water Storage in a Post-Fire Tundra Landscape

BRANDEN WALKER[1], BRAMPTON DAKIN[1], ARNAB SINGH[1], AND PHILIP MARSH[1]

ABSTRACT

Tundra landscapes are characterized by heterogenous snow cover resulting from blowing snow transportation and deposition. Blowing snow is captured by vegetation as well as along topographic gradients resulting in large drift formations. It is these areas that contain a proportionally large amount of snow water storage relative to the surrounding open tundra. In 2023, a tundra fire (EV014-23) burned approximately 356 km^2 at the northern edge of the tundra-taiga transition zone. This fire burned much of the vegetation that would typically capture blowing snow and result in snow drift accumulation. In this study, we set out to document snow cover conditions within the tundra burn site to understand the impacts of tundra fires on snow accumulation and ablation. In September 2023, drone mapping captured the extent and severity of the burn across the study area only weeks after the initial burn. In April 2024, snow depth, density, water equivalent and snow-covered area were documented using a combination of *in situ* observations and drone remote sensing techniques. Snow pits were conducted to understand the snowpack microstructure and to identify the entrapment of wind-blown soot, which would effectively reduce the snowpack albedo during the snowmelt in May. To further understand the impacts of tundra fires on snow conditions, results from the burn were compared to adjacent unburned regions as well as at multiple tundra sites located at the nearby snow conditions at the Trail Valley Creek Research Watershed, NWT.

[1] Cold Regions Research Centre, Wilfrid Laurier University. Waterloo, ON, Canada
Corresponding author: bwalker@wlu.ca

Deriving High-Resolution Snow Depth Maps and Soil Roughness Parameters from Unpiloted Aerial Systems (UAS) Equipped with Lidar for Modeling and Validating Airborne Radar Ku- and L-Band Backscatter Contributions

PETER TOOSE[1], BENOIT MONTPETIT[1], JULIEN MELOCHE[1], ALEX GÉLINAS[2], MICHAEL DAVEY[1], ARVIDS SILIS[1], CHRIS DERKSEN[1], AND RICHARD KELLY[3]

ABSTRACT

An Unpiloted Aerial Systems (UAS) was flown over 93 hectares of mixed pasture and woodlands near Powassan, ON, Canada, to collect coincident ground validation measurements for the University of Waterloo's high resolution airborne Ku- and L-band Cryosphere-Observing SAR (CryoSAR) system. Measurements of radar backscatter are sensitive to both snow and surface roughness properties and UAS equipped with lidar and cameras have been shown to be effective methods for measuring continuous snow depths at fine spatial and temporal resolutions and retrieving soil roughness statistics. In this paper we present high resolution snow depths maps produced by differencing digital elevation models created from UAS lidar point clouds, creating a time series of snow depth maps capturing three periods of sequential snow accumulation and one period of ablation (median snow depths of 0.16 m, 0.31 m, 0.48 m, 0.27 m), from January through March, during the 2022/2023 winter season. In-addition, the UAS was flown lower and slower producing point clouds with higher point densities over select sub-regions within the study domain where in-situ surface-based roughness measurements using mobile phone structure from motion (SfM) photogrammetry techniques were recorded. Here we provide preliminary results comparing soil roughness statistics derived from multiple sources at varying resolutions to better predict their influence on the airborne radar backscatter and subsequent radar retrieval of snow information.

[1] Climate Research Division, Environment and Climate Change Canada

[2] Département des Sciences de l'Environnement, Université du Québec à Trois-Rivières, Trois-Rivières, QC, Canada

[3] Geography and Environmental Management, University of Waterloo, Waterloo, ON, Canada

Corresponding author: peter.toose@ec.gc.ca

SNO-FOO 2024 – Samuel Tuttle

The 2024 SNO-FOO award is presented to Samuel Tuttle, faculty at Syracuse University and incoming Eastern Snow Conference Vice-President. May this year's award be a reminder that students are always watching.

Nothing is ever expected in the field with Sam. Whether it was watching him struggle putting on cable glands because he had to buy the wrong ones due to supply issues or trekking a muddy trail in northeastern Vermont to find your instruments chewed up by beavers. While all funny (mostly), nothing probably beats our most memorable trip to the Tenderfoot Creek Experimental Forest in Montana in late February and early March.

Why was it so memorable? For one, I got to ride a snow mobile (ok, I clung onto Dr. Eric Sproles' back the entire time), but it was also extremely challenging trip. Back to the story: Before we set off, Eric clearly warned us "Anything important goes into a zipped pocket". After a long cold night at the cabin with no power, we woke up and prepared to set off. Eric went to prepare the snow mobiles, Sam went to dig out the gate, and I cleaned up the cabin. As we were ready 2 hours later, I saw Sam – with a growing worrying expression on his face - going through his pockets: pants, parka, bags. It started to dawn on Sam that he had probably lost his phone in the snow. Searching the cabin turned up empty handed. The only place it could be was where he was digging. So, as we headed out to the main road (finding that Eric had gotten one of the snow mobiles stuck), Sam started the desperate dig to find his phone. As I left Sam to dig for his phone to see what help I can offer Eric, I hear the loudest shout of joy and relief. Sam found his phone and we were ready to set off and get the snow mobiles stuck like 3 more times that day.

Past Recipients

The Eastern Snow Conference annually bestows upon a distinguished snow scientist who, in striving for excellence in snow research, contributes to an event of notable humor the highly coveted Sno-Foo Award.

Year	Winner	Affiliation
2024	Samuel Tuttle	Syracuse University
2023	Peter Toose	Environment and Climate Change Canada
2019	Alexandre Roy	Université du Québec à Trois-Rivières
2018	Alexandre Langlois	Université de Sherbrooke
2017	Michael Durand	The Ohio State University
2016	Roger De Roo	University of Michigan
2015	Kevin Côté	Université de Sherbrooke
2014	Dorothy Hall	NASA GSFC
2013	Benoit Montpetit	Université de Sherbrooke
2012	Don Pierson	NYC DEP
2011	Ken Rancourt	Mt. Washington Observatory
2010	Kyung-Kuk (Kevin) Kang	University of Waterloo
2009	Rob Hellström	Bridgewater State University
2008	Steven Fassnacht	Colorado State University
2007	The Group of 9[1]	
2006	Andrew Klein	Texas A&M University
2005	Claude Duguay	University of Alaska-Fairbanks
2004	Chris Derksen	Meteorological Service of Canada
2003	Miles Ecclestone	Trent University
2002	Danny Marks	U.S.D.A.
2001	Brenda Toth	University of Saskatchewan
2000	Mauri Pelto	Nichols College
1999	Ross Brown	Meteorological Service of Canada
1998	Mary Albert	CRREL
1997	Jean Stein	Université de Laval
1996	Colin Taylor	Trent University
1995	Mike Demuth	NHRI
1994	Bert Davis	CRREL
1993	John Pomeroy	NHRI
1992	Tom Niziol	National Weather Service
1991	Terry Prowse	NHRI
1990	Kersi Davar	University of New Brunswick
1989	Gerry Jones	INRS-EAU
1988	Robert Sykes	SUNY
1987	John Metcalfe	Meteorological Service of Canada
1986	Peter Adams	Trent University
1985	Don Wiesnet	National Weather Service
1984	Barry Goodison	Meteorological Service of Canada

[1] Jimmy MacDonald (University of Saskatchewan), Bill Floyd (University of British Columbia), Chris DeBeer (University of Saskatchewan), Wendell Koning (Government of Alberta), Jaime Hood (University of Calgary), Dankia Muir (University of Calgary), John Jackson (University of Calgary), Sarah Forte (University of Calgary), Masaki Hayashi (University of Calgary)